World Scientific Series in Digital Forensics and Cybersecurity - Vol.3

A Practical Guide on Security and Privacy in Cyber-Physical Systems

Foundations, Applications and Limitations

World Scientific Series in Digital Forensics and Cybersecurity

Print ISSN: 2661-4278
Online ISSN: 2661-4286

Series Editor: Sanjay Goel, *The State University of New York at Albany*

This book series covers the latest research in the field of digital forensics as well as the state-of-the-art practice in the field. Eminent researchers and practitioners have been selected to work on different volumes of the series that will be announced and released in a sequence.

Published

Vol. 3   *A Practical Guide on Security and Privacy in Cyber-Physical Systems: Foundations, Applications and Limitations*
edited by Prinkle Sharma

Vol. 2   *Innovations in Digital Forensics*
edited by Suryadipta Majumdar, Paria Shirani and Lingyu Wang

Vol. 1   *SecureCSocial: Secure Cloud-Based Social Network*
by Pradeep K Atrey and Kasun Senevirathna

World Scientific Series in Digital Forensics and Cybersecurity - Vol.3

A Practical Guide on Security and Privacy in Cyber-Physical Systems

Foundations, Applications and Limitations

Editors

Prinkle Sharma

Sanjay Goel

University of Albany – The State University of New York, USA

W⊖ World Scientific

NEW JERSEY · LONDON · SINGAPORE · BEIJING · SHANGHAI · HONG KONG · TAIPEI · CHENNAI · TOKYO

Published by

World Scientific Publishing Co. Pte. Ltd.
5 Toh Tuck Link, Singapore 596224
USA office: 27 Warren Street, Suite 401-402, Hackensack, NJ 07601
UK office: 57 Shelton Street, Covent Garden, London WC2H 9HE

Library of Congress Cataloging-in-Publication Data
Names: Sharma, Prinkle, editor. | Goel, Sanjay (Of University of Albany, SUNY), editor.
Title: A practical guide on security and privacy in cyber-physical systems :
    foundations, applications and limitations / editors, Prinkle Sharma, Sanjay Goel,
    University of Albany--The State University of New York, USA.
Description: Hackensack, NJ : World Scientific, [2024] | Series: World Scientific series in digital
    forensics and cybersecurity, 2661-4278 ; vol. 3 | Includes bibliographical references and index.
Identifiers: LCCN 2023016742 | ISBN 9789811273544 (hardcover) |
    ISBN 9789811273551 (ebook for institutions) | ISBN 9789811273568 (ebook for individual)
Subjects: LCSH: Cooperating objects (Computer systems)--Security measures. |
    Internet of things--Security measures.
Classification: LCC QA76.9.A25 P733 2024 | DDC 006.2/2--dc23/eng/20230905
LC record available at https://lccn.loc.gov/2023016742

British Library Cataloguing-in-Publication Data
A catalogue record for this book is available from the British Library.

For any available supplementary material, please visit
https://www.worldscientific.com/worldscibooks/10.1142/13337#t=suppl

Desk Editors: Aanand Jayaraman/Amanda Yun

Typeset by Stallion Press
Email: enquiries@stallionpress.com

About the Editors

Prinkle Sharma is Assistant Professor at the Information Security and Digital Forensics Department at the University at Albany, State University of New York. Dr. Sharma received a PhD (May 2020) in Computer Engineering from the University of Massachusetts Dartmouth, specializing in cybersecurity networks and artificial intelligence. She has been working on securing wireless communications in autonomous vehicles by applying Artificial Intelligence since 2015 and has conducted extensive research in detecting security vulnerabilities in automotive systems. Her work on connected vehicle safety was featured in the IEEE Xplore Innovation Spotlight in 2018. Her research interests lie in network security, artificial intelligence, human–computer interaction, and machine learning.

Sanjay Goel is Professor and Chair of the Information Security and Digital Forensics Department at the School of Business and the Director of the NY State Center for Information Forensics and Assurance at the University at Albany, State University of New York (UAlbany), USA. He is also the Director of the Digital Forensics BS and MS Programs at the UAlbany, which he started. Dr. Goel received his PhD in Mechanical Engineering from Rensselaer Polytechnical Institute, Troy, New York, USA, and his MS in Mechanical Engineering from Rutgers University, Piscataway, New Jersey, USA. His research interests include information security, cyberwarfare, music piracy, complex systems, security behavior, and cyber-physical systems. His research on self-organizing systems includes traffic light coordination, smart grids, and social networks. He is actively engaged in policy efforts on cybersecurity norms, confidence-building measures (CBMs), and cyber treaties. He won the promising Inventor's Award in 2005 from the SUNY Research Foundation. In 2006, he was awarded the SUNY Chancellor's Award for Excellence in Teaching, the UAlbany Excellence in Teaching Award, and the Graduate Student Organization Award for Faculty Mentoring. In 2010, he was awarded the UAlbany Excellence in Research Award. In 2015, he was also awarded the SUNY Chancellor's Excellence in Academic Service Award, UAlbany President's Excellence in University Service Award, and School of Business Excellence in Research Award. He was named one of the three AT&T Industrial Ecology Faculty Fellows for 2009–2010. He has received grant funding totaling over 10 million dollars from multiple sources including the National Institute of Justice, the U.S. Department of Education, the US Department of Commerce, the National Science Foundation, the Intelligence Advanced Research Project Activity, Region II University Transportation Research Center, New York State Energy Research and Development Agency (NYSERDA), Blackstone Foundation, AT&T Foundation, and James S McDonnell Foundation and Blackstone Foundation. He has published over 100 articles in refereed journals and conference publications including top journals. He is a recognized international expert in information security, cyber warfare, and smart grid and has given plenary talks at events across several countries. In addition, he has been invited to present at 50 conferences including over 15 keynotes and plenary talks. He established the Annual Symposium on Information Assurance and the International Conference on Digital Forensics and Cyber Crime (ICDF2C).

About the Contributors

Nibedita Adhikari is attached to the PG Department of Computer Science and Applications, Utkal University, Bhubaneswar, Odisha, India. She holds an MSc in Math, MCA, and MTech (CSE) from the National Institute of Technology, Rourkela, India. She has also received her PhD in Computer Engineering from Sambalpur University, India. In her 20 years of teaching experience, she has guided many research scholars at the Postgraduate level and published 40 journal and conference papers. Her research interests include computer architecture, parallel and distributed processing, soft computing, and big data.

Sayada Sonia Akter received her BSc degree in Electrical and Electronic Engineering in 2016 and her MSc degree in Computer Science and Engineering in 2022, both from United International University, Dhaka, Bangladesh. She currently works as a Network Engineer at a financial institution in Bangladesh. Her research interests include software defined networking, network security, cryptography, and resource-efficient protocol for data aggregation in wireless sensor networks.

Pradeepkumar Bhale received his BEng in Computer Science and Engineering from the Government College of Engineering, Aurangabad, Maharashtra, India, in 2010. Following that, he worked with Tech Mahindra Pvt. Ltd. as a Technical Associate for a period of 2 years. He later completed his MTech in Information Security at Atal Bihari Vajpayee Indian Institute of Information Technology and Management, Gwalior, Madhya Pradesh, India, in 2014. He was awarded the State of Maharashtra's Postgraduate Fellowship to pursue his MTech degree. After that, he joined Dr. BR Ambedkar National Institute of Technology, Jalandhar, Punjab, India, as an Assistant Professor for 2 years. He is currently pursuing his PhD at the Department of Computer Science and Engineering at the Indian Institute of Technology Guwahati, Assam, under the supervision of Prof. Sukumar Nandi and Prof. Santosh Biswas. His current research interests include the Internet of Things, wireless security, network security, and discrete event system modeling.

Chidera Biringa received his Bachelor's (2017) and Master's (2021) degrees in Computer Science from Bells University of Technology, Ota, Nigeria, and the University of Massachusetts Dartmouth (UMassD), USA, respectively. He is currently pursuing a PhD in Engineering and Applied Science at UMassD, where he is a Research Assistant at the Cybersecurity Laboratory, NSA/DHS Designated Center of Academic Excellence in Cyber Defense Research (CAE-R). His research interests include program analysis for handling software vulnerability and machine learning for microarchitectural security.

Amardeep Das received his BEng in Computer Science Engineering and MTech degrees in Information Technology from Biju Patnaik University of Technology (BPUT), Odisha, India. He is currently pursuing his PhD at the Department of Computer Science and Applications at Utkal University, Bhubaneswar, India. He is also Assistant Professor at the Department of Computer Science and Engineering at CV Raman Global University, Bhubaneswar, India. He has more than 16 years of teaching experience. His research interests include the Internet of Things (IoT), intrusion detection systems (IDS), cyberphysical systems (CPS), security and privacy in wireless communication, cloud computing, mobile computing, and network security. He has published many papers in Scopus and SCI-indexed journals, conference papers, and book chapters.

Gökhan Kul is Assistant Professor at the Department of Computer and Information Science and the Associate Director of the Cybersecurity Center at the University of Massachusetts Dartmouth (UMass Dartmouth), USA. Before joining UMass Dartmouth, he was Assistant Professor at Delaware State University, USA. His research broadly covers software and systems security. He has published in reputable journals and at conferences such as IEEE TKDE and ACM TMIS focusing on data leakage, concept drift, and threat detection.

He contributes to research reproducibility efforts at VLDB and SIG-MOD reproducibility committees. He received his PhD in August 2018 at the University at Buffalo, USA, where he also served as an adjunct instructor for two semesters. He received his MS degree at Middle East Technical University (METU), Turkey, while working as a software engineer at the METU Computer Center.

Hong Liu is an IEEE senior member and Professor of Electrical and Computer Engineering (ECE) at the University of Massachusetts Dartmouth (UMass Dartmouth), USA. She has held other academic positions such as Academic Associate (full-time non-tenure-track faculty) for three years (1987–1989) at the School of Engineering, New York University, USA, and Visiting Scholar for two summers in 1992 and 1993 at Columbia University, USA. Dr. Liu received her BS degree (with Honors) in Computer Science and Mathematics (dual-major) and MS degree in Computer Science from Hefei Polytechnic University, the People's Republic of China, in 1982 and 1984, respectively. She received a PhD in Computer Science from New York University, USA, in 1990.

Dr. Liu integrates research, education, and application in computer networks/Internet of Things (IoT), cyber-physical systems (CPS), programming languages/compilers, machine learning, and network security. She has published numerous papers with her students and collaborators in refereed journals and peer-reviewed conference proceedings, and her publications for the past two decades have focused on cybersecurity. Dr. Liu serves at the MA Cybersecurity Education and Training Consortium (CETC).

Her research has been supported by USA's National Science Foundation (NSF), Department of Defense (DOD), Department of Transportation (DOT), Commonwealth Information Technology

Initiative (CITI), Massachusetts Information Turnpike Initiative (MITI), New York State Telecommunications Association (NYSTA), and Westinghouse Research Grant for Women in Electrical Engineering and Computer Science.

Sean McBride is Program Director of Industrial Cybersecurity at Idaho State University, USA. He runs USA's only hands-on degree to specialize in defending industrial facilities from cyberattacks and incidents. Sean holds an MBA from Idaho State University (2006), a Master in Global Management from Thunderbird School of Management, Arizona State University, USA (2010), and a PhD from La Trobe University, Australia (2021). His previous professional positions include Director of Industrial Control Systems Security at FireEye, Director of Analysis at Critical Intelligence, and Cybersecurity Researcher/Analyst at Idaho National Laboratory.

Arindam Pal is Senior Research Scientist at Data61 at the Commonwealth Scientific and Industrial Research Organisation (CSIRO), Australia, and Conjoint Senior Lecturer at the School of Computer Science and Engineering at the University of New South Wales (UNSW Sydney), Australia. His research interests lie in algorithms, artificial intelligence, data science, optimization, and machine learning. He works on business and research problems at CSIRO and collaborates with faculty members of universities both in Australia and abroad. He earned his PhD in Computer Science from the Indian Institute of Technology Delhi, India. He has over 15 years of research experience with software companies like Microsoft, Yahoo!, and Novell. He has published academic papers in reputed conferences and journals and was granted patents in India, USA, and Europe. He is a technical program member for several reputed conferences and a technical reviewer for many renowned journals. He is a Senior Member of both ACM and IEEE.

Mohammad Shahriar Rahman is currently Professor of Computer Science and Engineering at United International University, Bangladesh. He received his PhD and MSc in Information Science from Japan Advanced Institute of Science and Technology (JAIST), Japan, in 2012 and 2009, respectively, and a BSc in Computer Science and Engineering from the University of Dhaka, Bangladesh, in 2006. He has worked as a research engineer at the Information Security group of KDDI Research, Japan. His research interests include secure protocol construction, privacy-preserving computation, and security modeling to solve problems of an increasingly knowledge-based connected world facilitated by blockchains, smart grid, smart cities, the Internet of Things (IoT), and cloud computing. He is a member of the International Association for Cryptologic Research (IACR) and the Institute of Electrical and Electronics Engineers (IEEE). Prof. Rahman has co-authored 60+ research papers and submitted 8 co-authored Japanese patent applications.

Sushmita Ruj is Senior Lecturer at the School of Computer Science and Engineering, University of New South Wales, Sydney, Australia. Her research interests lie in blockchains, applied cryptography, and data privacy. She received her BE degree in Computer Science from the Indian Institute of Engineering Science and Technology (IIEST), Shibpur, India, and a Master's and PhD in Computer Science from the

Indian Statistical Institute. She was previously Senior Research Scientist at Data61, Commonwealth Scientific and Industrial Research Organisation (CSIRO), Australia, and Associate Professor at the Indian Statistical Institute, Kolkata, India. She serves as Associate Editor of *IEEE Transactions on Information Forensics and Security*, *Information Security and Applications*, and *Pervasive and Mobile Computing*. She is a senior member of the ACM and IEEE. She is a recipient of the Samsung GRO award, NetApp Faculty Fellowship, Cisco Academic Grant, and IBM OCSP grant.

Corey Schou is Professor of Cyber Security and Director of the Informatics Research Institute and the National Information Assurance Training and Education Center at Idaho State University, Pocatello, Idaho, USA. Corey received his undergraduate degree in BIO/Chem at Rollins College, Winter Park, Florida, USA. He has a Master's degree in Oceanography and a PhD from Florida State University in International Law, USA. His research has been characterized as interdisciplinary and has focused on the emerging changes as cybersecurity has evolved from COMSEC. He has written 3 books, 300+ papers and presentations, as well as dozens of reports and position papers for the US Government. His body of work was recognized in 2019 by his induction into the Cyber Security Hall of Fame. In addition, he was the first recipient of the (ISC)2 Tipton Award, FISSEA Educator of the Year, and Fulbright Scholar (New Zealand). He was one of the founders of both the Colloquium for Information Systems Security Education (CISSE) and the International Information System Security Certification Consortium. In his role as Senior Advisor to the US Department of State, he has worked with over 60 governments by helping them develop a compendium of standards for cybersecurity.

Jill Slay is currently the University of South Australia SmartSat Professorial Chair in Cybersecurity and Research at the Smart-Sat Australian Co-operative Research Centre (CRC). Jill received her Engineering degree from the University of Herts, UK, and her PhD from the Curtin University of Technology in Perth, Western Australia. Her work focuses on the context of helping to develop a national technical agenda in satellite cybersecurity and resilience within the Australian Defence Industry. She is ranked as being in the top 2% of scientists in the world in the ICT Networking and Telecommunications subfield (in 2019) as an early adopter of AI and machine learning in cybersecurity and real-time forensics. She applies these techniques to satellite security. Her previous appointments have included Optus Chair in Cybersecurity at La Trobe University, Australia, and Founding Director of the Australian Centre for Cyber Security at the Australian Defence Force Academy.

Acknowledgment

Many people have earned my gratitude for their contribution to my time while working on this book. First, I am immensely grateful to all the authors for contributing their world-class research to this book. I would also like to acknowledge the reviewers for their timely and insightful feedback on the chapters.

Special thanks to my co-editor, Dr. Sanjay Goel, my colleagues, and friends with whom I have had the pleasure to work during this project. As a result of your efforts and encouragement, I was able to turn this dream of editing a book into reality.

Furthermore, I owe an enormous debt of thankfulness to the World Scientific Publishers' team, who gave me their time to discuss early drafts and provided timely guidance throughout the process.

Finally, a huge thanks to my Mom and Dad. Without your unconditional support, love, trust, motivation, and continued patience, I would not have been able to work effectively on this book project.

Contents

Introduction

The Internet has seen transformation several times since its inception — from the introduction of various forms of electronic communication to e-commerce, social media, and cyber-physical systems (CPSs); each has deeply impacted society. The most recent development incorporates CPSs where sensors and actuators enable the Internet to impact the physical world. Three major innovations driving technologies all rest on CPSs: smart power grids, smart wearable/implantable/medical devices, and connected/autonomous vehicles.

While each of these innovators can single-handedly transform society, security and privacy concerns still loom overhead and need to be addressed adequately to ensure the mobility of these systems. Because CPSs have real-world impacts on the physical world through the Internet, hackers can exploit this connection to cause damage to life and property. Just imagine a ransomware attack where the hacker has control of the oxygen supply of a hospital in a way that grants them the ability to kill patients if the ransom demand is not met. Another scenario is where a hacker can gain access to the vehicles in a transportation network and disable the self-driving cars' collision avoidance systems, resulting in numerous accidents on the highway. Both examples show that the Internet's impact on society is growing exponentially, as is the potential for damage through attacks.

Consumer privacy faces an increasing risk as we collect more and more data from consumers, for instance, physiological data through wearables. Data have become a commodity of immense value primarily for marketing and consumer research. However, they have also been used for less benevolent purposes; some firms are leveraging data for financial gain at the cost of consumer privacy. While legislation on privacy (e.g., the EU's General Data Protection Regulation (GDPR)) is being adopted by governments, the risk of a breach in and/or abuse of consumer privacy remains — with some fallouts being potentially detrimental to consumers.

A Practical Guide on Security and Privacy in Cyber-Physical Systems offers an in-depth look at CPSs' recent security and privacy challenges in several application domains. The focus is on the foundational components and architecture of CPSs, the network and communication protocols, and several case studies highlighting the security and privacy challenges of CPSs followed by studying the forensics of these systems. This book further provides a systematic overview of several CPS applications by including an insight into the standard architectures developed. Then, it delves into each of the layers of such architectures and describes the underlying technological, security, and privacy issues that are currently faced by some of the main CPS research groups. A comprehensive vision of CPS applications in smart vehicles, smart grids, and energy systems is also presented. Finally, the book offers the guiding principles that should be observed while planning future innovations for such systems by highlighting the security gap within existing protocols.

This book is divided into three parts — Part 1: Foundations; Part 2: Applications; and Part 3: Limitations — to provide the reader with a deeper understanding of CPSs and their operations by defining **what** CPSs are, **how** they operate, and **why** exactly they are needed. The three parts of the book are independent of one another and are meant to be read concurrently.

CPS technology is constantly evolving, and this book captures the latest advancements from multiple aspects. The following are summaries of each chapter in this book and the topics they cover:

Chapter 1 covers the basic components and computing concepts required for structuring any cyber-physical system. It identifies the key requirements for developing CPS architecture from initial requirements to certification and deployment. The chapter also explains CPSs' components, their properties, and the computing constraints of the physical world. It also analyzes the engineering required for CPSs, including the application of sensing, actuation, control, communication, and software design.

Chapter 2 presents the various wireless MAC protocols and techniques for achieving real-time and reliable communications in CPSs in detail. It also discusses the protocols and requirements needed at the networking layers for these applications. It evaluates the IEEE standard 802.15.4 for configuring and optimizing the design parameters of the MAC protocols. Subsequently, it identifies the important parameters such as bandwidth, delay, reliability, security, and mobility, which are essential to allow for the effective and robust operation of the various CPSs.

Chapter 3 discusses the forensic analysis and tools required to investigate incidents when security measures fail to safeguard the CPSs. It covers the technical and legal aspects of digital forensics in CPSs and reviews current challenges that need to be addressed by the developers of CPSs. The chapter further highlights the need for opting for the forensic-by-design approach for different CPSs and defines the factors that contribute to building such frameworks.

Chapter 4 highlights another application domain of CPSs — cloud computing — which is proving to be a strong solution to an organization's and individual's ever-increasing storage and processing needs, without the hassle of owning and managing physical hardware. In this chapter, the authors review the cloud environment, the threats and attacks it may face, the approaches that cloud forensics may adopt for various frameworks, and the practical challenges and limitations in dealing with cloud forensic investigators.

Chapter 5 provides a detailed review of various vulnerability challenges in smart cities and a foundation for classifying current and future advancements in this field. In addition, it describes the security requirements for designing a security solution for a smart city, identifies the existing security solutions, and discusses the open research issues and challenges of smart city security.

Chapter 6 examines the self-healing capability of a smart grid system after a power failure. The work focuses on the potential cyberattacks that could disrupt the functioning of the CPSs. The core components including the grid assets and their classification are discussed in detail, providing the reader with an outlook on its vulnerable points to cyberattacks. Further, the chapter shows readers how cyberattacks may be classified based on protocols and their components since the ability to analyze attacks will help to protect smart grids against similar attacks in the future as well as avoid failures during disasters. Besides increasing the security level of the smart grid, smart grid forensics aids data and evidence collection, and assists in identifying data thieves and hackers. Lastly, the chapter introduces the emerging areas of smart grid forensics, discusses the challenges, and outlines the open issues in this topic.

Chapter 7 analyzes the existing Industrial Control Systems (ICS) security curricular guidance to explore its potential in information technology (IT) and operational technology (OT) within CPSs. The authors explore several scenarios demonstrating the ineffectiveness of ICS in bridging the cybersecurity gap within IT–OT systems and propose an updated knowledge unit. The proposed knowledge unit focuses on the importance of formalized education, and its effectiveness has been validated using various techniques in this chapter. Results from the newly developed knowledge unit are presented in a survey to guide cybersecurity professionals who can now comfortably interact with both IT and OT systems without putting security at risk.

The book is a well-balanced combination of academic contributions and industrial applications in CPSs. It is written for students and professionals at all levels to provide the best practice for individuals who want to advance their research and development in this exciting area by learning about state-of-the-art technologies.

https://doi.org/10.1142/9789811273551_0001

Chapter 1

Cyber-Physical Systems and the Internet of Things: An Overview

Amardeep Das*,†,§, **Pradeepkumar Bhale**‡,¶,
and Nibedita Adhikari*,‖

**Utkal University, Vanivihar, Bhubaneswar, Odisha, India*
†*C.V. Raman Global University, Bhubaneswar, Odisha, India*
‡*Indian Institute of Technology Guwahati,
Guwahati, Assam, India*

§*amardeepcvrp@gmail.com*
¶*pradeepkumar@iitg.ac.in*
‖*nibedita.cs@utkaluniversity.ac.in*

Abstract: Cyber-physical systems (CPSs) more significantly impact the present and future of engineered systems. In the areas of design, implementation, and applications, the integration of CPSs with other technologies frequently presents distinctive difficulties. Our proposed chapter will look at different definitions of the integrated cyber-physical system in IoT platforms, field evolution, and new research areas. CPSs offer several implementation challenges related to efficiency, reliability, predictability, and security. When combined with the Internet of Things (IoT), CPSs develop into a hybrid technology that advances the technical aspects. The CPS–IoT models work best together because they provide helpful insights into dealing with engineering systems and control modules.

Keywords: Cyber-Physical Systems (CPSs), Internet of Things (IoT), Hybrid Technology, Security.

1.   Introduction to Cyber-Physical Systems and Internet of Things

Helen Gill of the National Science Foundation developed a cyber-physical system in 2006. Cyber-physical systems (CPSs) incorporate cyber software control and physical mechanism components. CPSs enhance communication seamlessly with physical and computational networks. These technologies will improve our infrastructure and promote a healthy lifestyle. CPS sensors and actuators coordinate and interact with computational elements [1,2]. They will enhance personalized healthcare, emergency response, traffic flow management, and power generation and delivery (cars, buildings, homes, cities, manufacturing, hospitals, appliances, and so on). The Anti-lock/ Anti-skid Braking System (ABS) in a car, for example, is a built-in system that controls the amount of force used to stop the vehicle. This computer system interacts with the natural world through sensors and actuators. These embedded systems are no longer independent. Instead, they share data over communication networks like the Internet, where data from several embedded systems can be collected and analyzed using cloud computing [3,4]. A computing unit can control and decentralize embedded systems that are connected. The information collected could be processed by hand or with the help of a Human–Machine Interface (HMI) [5].

A CPS uses IoT as its foundational or enabling technology. CPSs are the advancement in conception and perception of the Internet of Things (IoT), and they have a significant capacity approve for physical world control. Traditional embedded and control systems are also a part of CPSs, which have evolved into cutting-edge methodologies. For dependable transmission and information processing, IoT links information-acquiring devices like sensors, RFID (Radio Frequency Identification), wireless sensor networks, and cloud computing technology. In contrast, CPS is a control technology that combines computing, communication, and IoT control. It is scalable and reliable. While IoT, on the one hand, focuses on the transmission and processing of information, CPS, on the other hand, can not only sense but also has a potent ability to control. Cyber and physical components are linked to one another in CPS on both a spatial and temporal scale, revealing various distinct behavioral processes that interact with one another in various ways that change the context. Mechatronics, computation theory, the IoT, wireless sensor networks, theory of cybernetics, design, and process science are all combined in the transdisciplinary field of process science. Control is a function of embedded systems that employ feedback loops. The IoT and CPS have the same architecture, which makes them similar, but CPS places a higher priority on separating physical from computational elements. CPSs are portrayed explicitly as a structure of interacting elements with physical input and output rather than standalone devices, in contrast to traditional embedded systems [6]. Healthcare and industrial automation, control technology, distributed energy systems, aircraft control, and other areas are applications of CPSs [7]. This sentence implies that traditional engineering systems, which focus on the physical aspects of design and operation, will evolve into Cyber-Physical Systems (CPS). As a result, there will be improvements in economic welfare. The objective is to develop user-centric ecosystems comprising cutting-edge, viable technologies that apply IoT and CPS innovative solutions to increase efficiency by enabling unobtrusive, adaptable, and highly usable services at the network edge, gateway, and cloud levels [2,6,8].

The rest of this chapter is organized as follows: Section 2 describes the CPS characteristics and their application. Section 3 deals with the 5C Layer Architecture of CPS. Section 4 describes the overview of IoT and its layers briefly. Section 5 presents a comparative study of CPS and IoT architecture. Section 6 presents a CPS/IoT attack tree. In Section 7, open issues are summarized for further research. Finally, concluding remarks are presented in Section 8.

2. CPS Characteristics and Applications

The main characteristics of CPS are discussed in this section. The main three parameters that make CPS popular are (i) *Intelligence* — Adaptive and Robustness, (ii) *Network* — Communication, Cooperation, and Cloud solutions, and (iii) *User-friendliness* — CPS has computation and physical processes [1,2]. The Software is embedded into the physical systems, and CPS networks use wireless sensor networks. CPS can be used in many applications, including intelligent transportation, precision agriculture, Health CPS, and computer networks. The various domains are described briefly in the following.

2.1. *Vehicular CPS*

Future aircraft and air traffic control systems are affected by the development of CPS. Accidents caused by human error can be significantly reduced or even eliminated by CPS technologies. To avoid accidents or traffic jams, improve safety, and ultimately save money and time, individual vehicles and the infrastructure can communicate with one another and share real-time information about the flow of traffic, the location of problems, and other relevant information [9,10]. The vehicular cyber-physical system is shown in Figure 1.

2.2. *Agricultural CPS*

In agriculture, CPS is a crucial piece of technology for precise farming. The goal is to monitor soil moisture so that plowing can be planned, monitor soil mineral content so that fertilization can be planned, monitor the weather so that it does not freeze, monitor crop growth so that diseases and pests can be controlled, and manage yield. Full integration of CPS with smart devices in the IoT is considered a step toward autonomous control of all agricultural production processes, which assures stability and adaptability to environmental and market changes [12,13]. The agricultural CPS is shown in Figure 2.

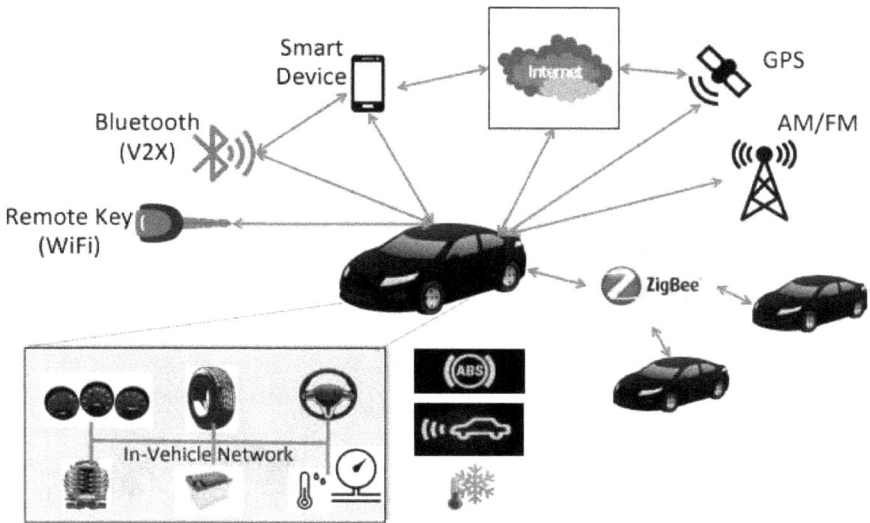

Figure 1. Network vehicular cyber-physical system [11].

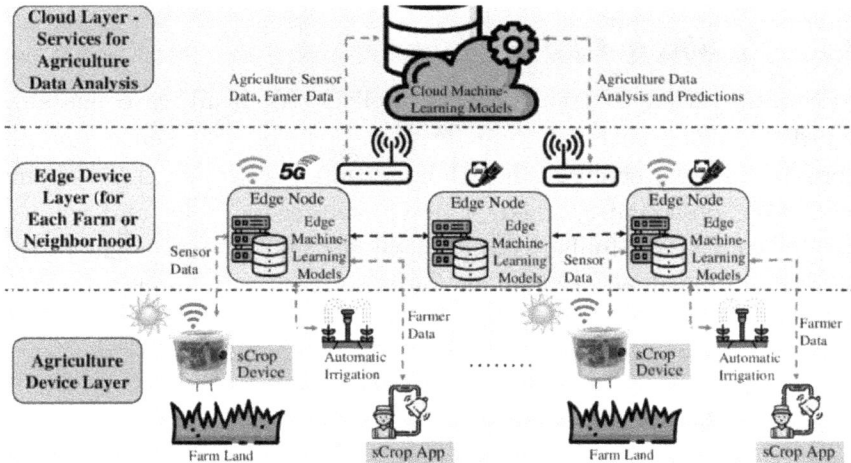

Figure 2. Agricultural CPS [14].

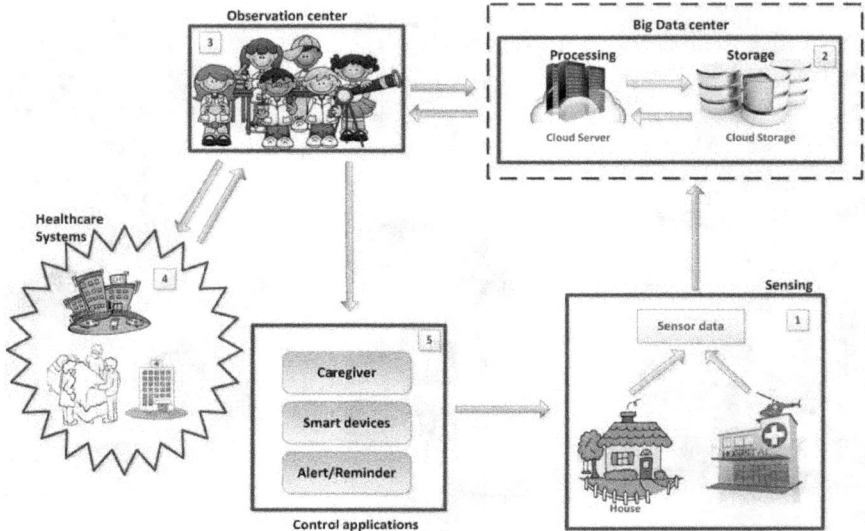

Figure 3. A CPS for healthcare monitoring [16].

2.3. *Healthcare CPS*

CPSs are poised to transform the delivery of healthcare enabling smart medical treatments and services. For example, sensors in the home will detect changing health conditions, new operating systems will make personalized medical devices interoperable, and robotic surgery and bionic limbs will help heal and restore movement to the injured and disabled and one day even augment human abilities [15]. The healthcare CPS is shown in Figure 3.

2.4. *Computer network CPS*

CPS can boost cyber environments to understand system and user behaviors better. The integration of CPS functionality into social networks can have a profound impact. Popular social networks and e-commerce websites store users' navigation information and analyze the information to predict interests and recommend friends or products [17]. In Figure 4, the computer network CPS is displayed.

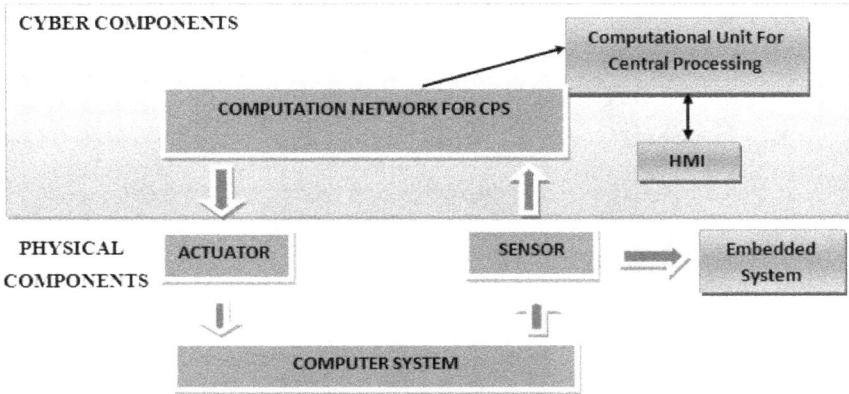

Figure 4. Block diagram of computer network CPS [17].

3. CPS 5C Level Architecture

The 5C architecture guides manufacturing CPS design and deployment. A CPS has two important functional components: (1) better connection that offers real-time data acquisition from the physical world and cyberspace feedback; (2) intelligent data management, analytics, and computational capabilities that construct cyberspace. This condition is too vague for widespread use. The 5C design outlines a sequential process for developing a CPS, from data collection to value production [18]. The CPS 5C level architecture, application, and methods are shown in Figures 5 and 6.

3.1. *Smart connection*

To build a cyber-physical system application, the first step is to gather accurate and reliable data from machines and their components. The information may come from enterprise manufacturing systems like ERP, MES, SCM, and CMM, controllers, or data that sensors have directly collected. At this level, two essential factors must be taken into account. Considering the various types of data, it is necessary to have a seamless and untethered method for managing the data acquisition procedure and transferring data to the central

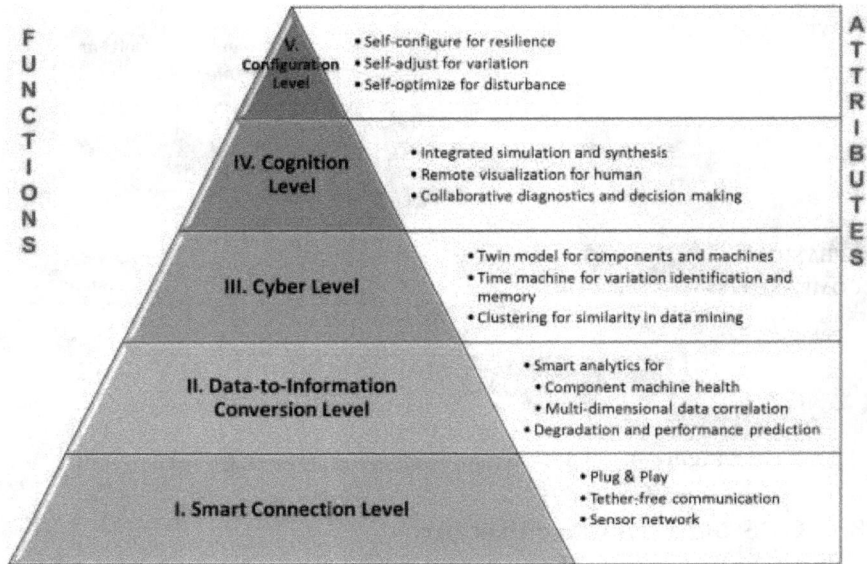

Figure 5. 5C architecture and implementation of CPS [19].

server, where specific protocols such as MT Connect and others are useful. The selection of appropriate sensors (type and specification) is the second important factor for the initial level.

3.2. Data-to-information conversion

Data must be analyzed to learn anything useful, and tools and methods that turn data into information are used. Many people are interested in using these algorithms to make predictions and care for health. The second level of CPS architecture gives machines self-awareness by letting them figure out their health value, how long they are likely to work, and so on [20].

3.3. Cyber

Cyber is the key information center in this architecture. Every connected machine sends its data, building a network. Massive volumes of data have been collected, so specific analytics must be employed to

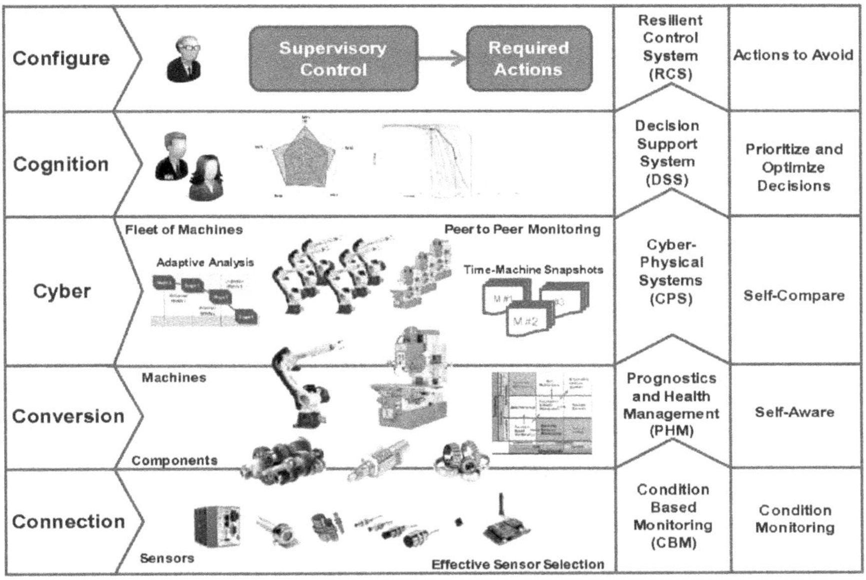

Figure 6. Applications and methods related to each 5C architectural level [18].

extract more data that increase fleet comprehension. These analytics give machines the ability to compare and rate their performance with that of the fleet, allowing for self-comparison. On the other hand, by comparing machine performance to earlier assets (historical data), it is possible to forecast how the machinery will behave in the future.

3.4. *Cognition*

This degree of CPS implementation provides a thorough insight into the monitored system. Properly presenting the knowledge to experts aids in decision-making. Using comparison, machine statuses and maintenance chores can be prioritized to maximum efficiency. This level requires infographics to convey knowledge to consumers.

3.5. *Configuration*

The configuration level serves as a supervisory control for machines to be self-configuring and self-adaptive. After making corrective and preventive decisions in cognition, the RCS (Resilience Control System) applies them to the monitored system.

4.   Internet of Things

IoT is a network of devices that can communicate and accomplish tasks without human or computer input. For intelligent recognition, location, tracking, monitoring, and management, the IoT involves putting sensors (RFID, IR, GPS, laser scanners, etc.) on everything and connecting them to the Internet through specific protocols for information exchange and communication [4,6].

4.1.   *Properties of Internet of Things*

A technological revolution known as IoT shows how computing and communication have integrated technology in the day-to-day life of every individual. Its growth depends on technological progress in many key areas, such as nanotechnology and wireless sensors. A few key points must be discussed, which must be in place for the proper functioning of IoT devices.

4.1.1.   *Connect both inanimate and living things*

For the first tests and industrial equipment, an IoT network is linked up. The IoT is now trying to connect everything, from everyday objects to big machines. Some of the things they make are gas turbines, cars, and utility meters. The expansive realm of the IoT holds the potential to embrace not only inanimate objects but also living entities such as plants, cherished pets, and even humans, enriching our understanding and care for these beings [7]. For example, the Cow Tracking Project in Essex uses information from radio positioning tags to keep an eye on how the herd is acting and see if any cows are sick. In the IoT world, people are connected through wearable computers and digital health devices like Nike+ FuelBand and Fitbit. Cisco has added to the definition of IoT, which includes people, places, and things. The new connected ecosystems can include anything to which you can add a sensor and connectivity.

4.1.2. *Use sensors for data collection*

One or more sensors will be present on the connected physical objects. Each sensor will monitor a particular factor, such as temperature, vibration, location, and motion. These sensors will be connected through IoT to other devices and programs that can interpret or display data from the sensor's feeds. A company's systems and people will receive new information from these sensors.

4.2. *Layer architecture of Internet of Things*

The IoT works in different layers, and each layer is responsible for performing toward the goal [21,22]. Figure 7 shows the layer architecture of IoT, which is based as proposed in [22].

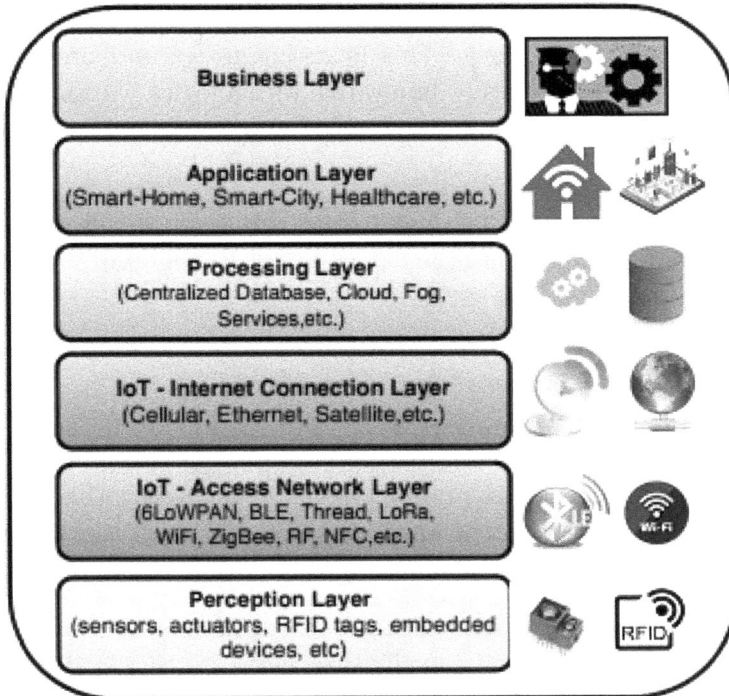

Figure 7. Layer architecture of Internet of Things [21].

Layer 1: Perception layer. This layer has a wireless identification and sensing platform (WISP), which is an RFID device like a sensor or actuator that supports sensing an object and computing: a microcontroller powered by radio-frequency energy [3,7].

Layer 2: Access network layer. In this layer, the different types of communication protocols like 6LoWPAN, BLE, Thread, LoRa, Wi-Fi, ZigBee, and NFC are incorporated to access the IoT device.

Layer 3: Internet connection layer. In this layer, the different types of networks like WPAN, WWAN, and WLAN communicate different types of information to one another.

Layer 4: Processing layer. This layer manages the information and its processes in different areas like data centers, search engines, smart decisions, information security, and data mining.

Layer 5: Application layer. This layer integrates and provides the different types of services to innovative technologies like smart logistics, smart grids, green buildings, smart transport, and environmental monitors.

Layer 6: Business layer. The last architecture layer includes the organization and management of IoT devices. It also maintains IoT application and profit models [22].

4.3. *Characteristics of Internet of Things*

The following are the IoT's core characteristics:

Intelligence: Algorithms produce the intelligent spark that provides product experience to intelligence computing (i.e., software and hardware) [23,24].

Connectivity: Connectivity in the IoT requires more than a Wi-Fi module [17]. Connectivity facilitates network compatibility and accessibility. Accessibility is the ability to connect to a network, whereas compatibility is the shared capacity to consume and produce data. If this sounds familiar, it is because Metcalfe's Law [25] is valid for the IoT.

Sensing: Sensing technology and its various applications constantly evolve in line with technological advancements and business needs [22]. Sensors can detect various real-world properties — from distance to heat to pressure. Today's products sense everything around them using sensors, and they can be highly accurate, consume less power, and are inexpensive to install and maintain [26]. As a result, sensors are vital in creating new value for their processes and respective businesses.

5. Comparison of CPS and IoT

This section presents the comparative study of CPSs and IoT based on the different parameters. It emphasizes making the application domains adaptable to a common goal of increasing technological reliability, expanding advancement opportunities, and exposing areas of untapped potential.

5.1. *Technology*

A CPS is smartly designed systems that combine computing, networking, and physical processes seamlessly. It is a system that combines physical and digital parts that could be connected to a network and work closely together. IoT is an umbrella term for the growing number of physical devices around the world that are connected to the Internet and, eventually, to each other. IoT is a networked world where devices, objects, and people are all linked to each other.

5.2. *Focus*

A CPS is a more general version of the idea of embedded systems. CPS represent an evolved concept of embedded systems, where the seamless fusion of cyberspace and physical processes allows for enhanced control and vigilant monitoring of real-world systems. In terms of how it works in real life, this integration considers how both the cyber and physical parts work together [24]. IoT, on the other hand, is a more advanced state where the digital and physical worlds are mixed into one space. It is about how these real things can be linked to the Internet to do something useful.

5.3.   *Mechanism*

Innovative embedded systems, such as cyber-physical ones, use sensor networks and embedded computing to keep tabs on their physical surroundings. Algorithms programmed into the mechanism keep tabs on and control it. As long as humans are involved in the process, these systems can assess operational conditions and assist in making decisions. The IoT, on the other hand, does not require any human intervention [15].

5.4.   *Scope*

IoT works on the level of physical things and focuses on making connections everywhere in the physical world. IoT is not just about connecting things; it also lets them talk to each other and share data, which can then be analyzed and turned into information that makes sense [7,20]. CPSs combine sensors or actuators with technologies for networking. Sensors and actuators work together in a feedback loop, where people can change their behavior based on what the user wants. IoT, on the other hand, originated from a separate set of groups, with CPSs being the primary focus (Figure 8).

A cyber-physical system combines computational and physical elements, software-based networking of mechanical and electrical systems. It employs process expertise to regulate logistics and production independently.

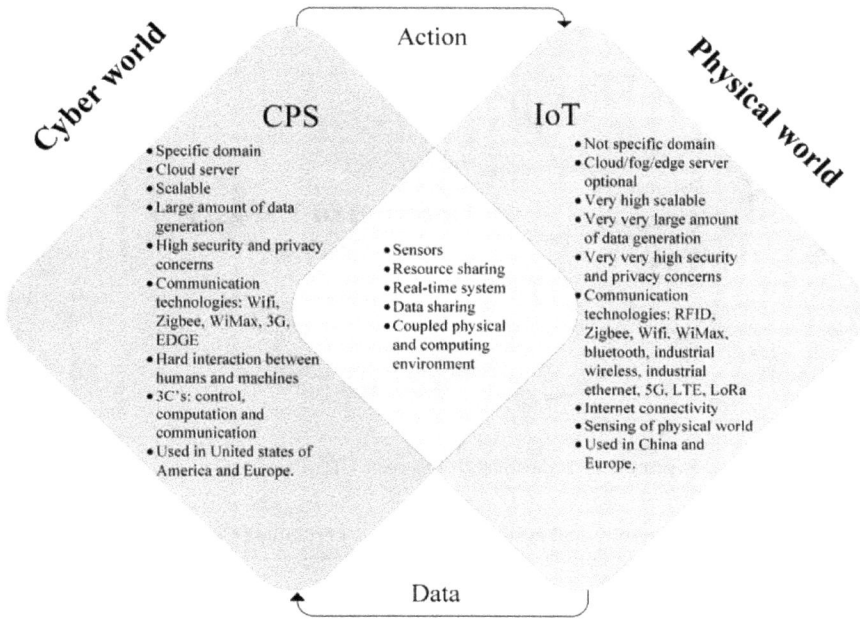

Cyber world

Action

CPS
- Specific domain
- Cloud server
- Scalable
- Large amount of data generation
- High security and privacy concerns
- Communication technologies: Wifi, Zigbee, WiMax, 3G, EDGE
- Hard interaction between humans and machines
- 3C's: control, computation and communication
- Used in United states of America and Europe.

- Sensors
- Resource sharing
- Real-time system
- Data sharing
- Coupled physical and computing environment

IoT
- Not specific domain
- Cloud/fog/edge server optional
- Very high scalable
- Very very large amount of data generation
- Very very high security and privacy concerns
- Communication technologies: RFID, Zigbee, Wifi, WiMax, bluetooth, industrial wireless, industrial ethernet, 5G, LTE, LoRa
- Internet connectivity
- Sensing of physical world
- Used in China and Europe.

Physical world

Data

Figure 8. Comparison of CPSs and IoT; supporting Industry 4.0 development [27].

On the other hand, the idea for the IoT came from networking and information technology, intended to bring the digital world into the real world [13,15,28]. IoT is a catch-all term for many things that involve getting the Internet and Web into the real world through the widespread use of devices with built-in identification, sensing, and action capabilities.

IoT aims to connect everything via the Internet. On the other hand, CPSs seek to connect everything with/without the Internet [1,15,20]. So, IoT is a subset of CPSs. As shown in Figure 9. The comparative analysis of CPSs and IoT is shown in Table 1.

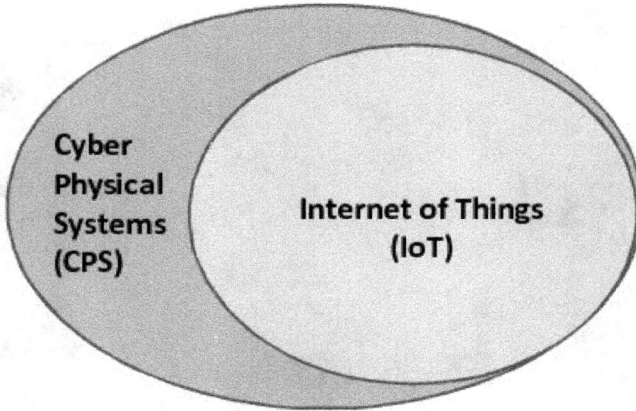

Figure 9. Relationship between CPSs and IoT [29].

Table 1. Cyber-physical systems vs Internet of Things: Comparison chart.

CPS	IoT
A system that combines physics with cyber components, potentially networked and tightly interconnected.	A catch-all term for the growing number of physical devices around the world that are connected to the Internet and to each other.
It seamlessly integrates computation, networking, and physical processes.	Comprises things or objects that have unique identities and are connected to the Internet.
Sensors and actuators work in the feedback loop using human intervention.	It is purely automation meaning no human assistance is required.

6. Tree of Attacks on CPS and IoT

The ISO/IEC 27001:2013 standard states that dangers might be intentional, unintentional, or environmental. The following are some examples of common threats: physical harm, natural disasters, interruption of vital services, radiation malfunctions, compromise of information, technical setbacks, unauthorized activities, and compromise of functions.

Based on an examination of existing security studies depicted in Figure 10, a "tree" of assaults and threats based on the functional

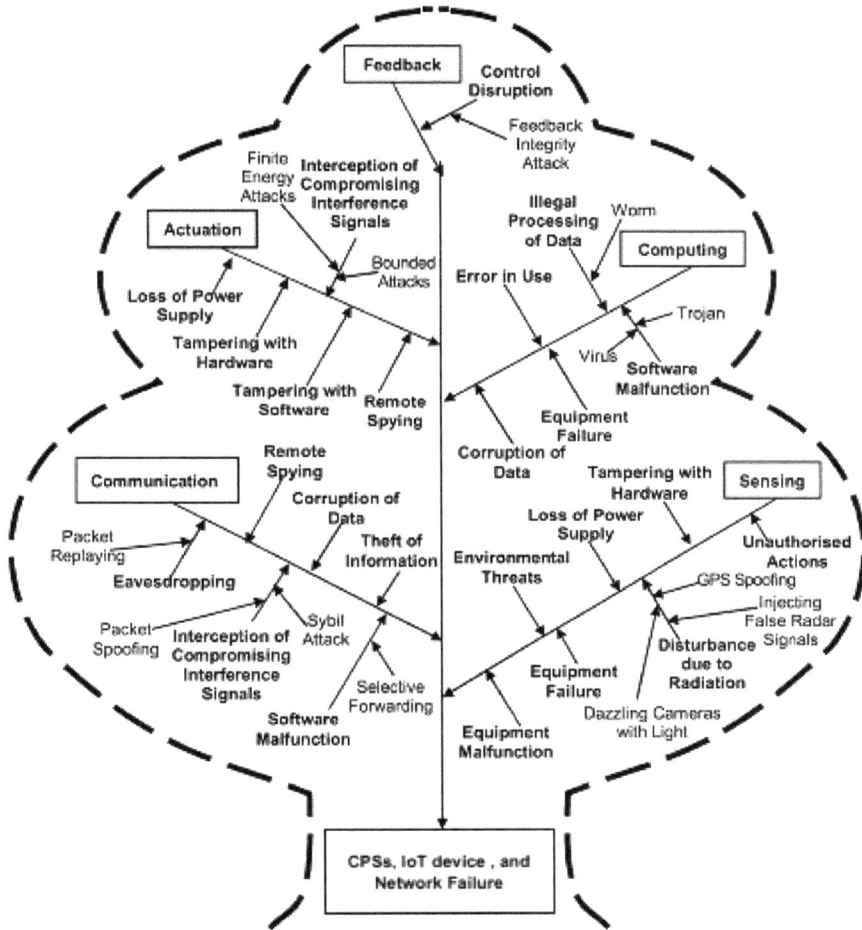

Figure 10. A tree diagram of attacks and perils on CPSs and IoT [30].

model of CPS and IoT is provided. The "tree" has branches that include the following five kinds of cyberattacks: (a) those that target sensors (sensing); (b) those that target actuators (actuation); (c) those that target computer components (computing); (d) those that target communications; and (e) those that target feedback (feedback).

7. Open Issues

CPSs and IoT can create new markets and solve societal problems, but they require good quality, safety, security, and privacy. Fundamental scientific research must counteract internal and external changes with predictable validation and measurement quality. Based on the above analysis of the most recent CPS and IoT security studies, the following tasks should be included in future research:

7.1. *CPS and IoT security protocol development*

Due to the large number of devices in CPSs and the IoT, there are many questions about how well and how easily modern security standards and protocols can protect the privacy and integrity of data. The use of intelligent security protocols, which allow the self-adoption and self-management of CPS architecture, and their incorporation into cutting-edge, inventive gadgets are among the most critical challenges. Interoperability problems might arise when different security technologies of different CPS components interact with one another. It takes a lot of work to make sure that the CPS as a whole has security and privacy features built-in for each of its parts.

7.2. *Development of countermeasures for vulnerabilities*

In order to reduce the number of vulnerabilities in the CPS, it is important to design countermeasures. It has become clear through analysis of current efforts to increase the dependability and resilience of CPSs that defensive mechanisms must be developed and their effects on CPS survival assessed.

7.3. *CPS and IoT security architecture*

Analysis of the primary CPS issues brought on by the expansion of fast-evolving physical and cyber risks demonstrates the necessity of establishing a fault-tolerant, dependable architecture that provides a high degree of security and efficiency.

7.4. *Security of personal data*

Due to the proliferation and improvement of digital technology, the safety of sensitive data is increasingly at risk. Data privacy may be violated due to unauthorized access to personal data. Malevolent users obtain sensitive information by analyzing data intelligently. This issue can be resolved using ethical and technological considerations. Transparency-based security is one of the alternatives.

7.5. *Security metric development*

According to cutting-edge research, depending on the risks involved, it is necessary to perform at a particular degree of trust while verifying the CPS's authenticity, secrecy, dependability, resilience, and integrity against various assaults. For comprehensive information, a CPS uses data from a number of sensors. The user may obtain inaccurate information as a result of a conflict between reliable operation in the event of a sensor failure and defective sensors.

7.6. *CPS and IoT authentication component development*

Using authentication procedures and a secure connection between sensors and controllers makes it harder to change the CPS.

8. Conclusion

CPSs combine physical and computational processes. CPSs have much greater economic potential than has been realized, and significant investments are being made all over the world to advance the technology. The capability, adaptability, scalability, resilience, safety, security, and usability of CPSs will far outpace those of today's straightforward embedded systems. CPSs have an effect on the users' personal, professional, and financial lives as a whole. In this chapter, a review was conducted on the definition of a CPS as well as its history, various fields of research, the integration of IoT and CPS, and how they complement each other as well as their applications.

The research fields of CPS and its advancements have been a part of a number of emerging trends in the information technology fields, such as the IoT, cloud computing, big data, Industrial Internet, and Industry 4.0.

References

[1] Tavčar, J. and Horvath, I. A review of the principles of designing smart cyber-physical systems for run-time adaptation: Learned lessons and open issues. *IEEE Transactions on Systems, Man, and Cybernetics: Systems*, 49(1):145–158, 2018.

[2] Humayed, A., *et al.* Cyber-physical systems security — A survey. *IEEE Internet of Things Journal*, 4(6):1802–1831, 2017.

[3] Zhang, Y., *et al.* Health-CPS: Healthcare cyber-physical system assisted by cloud and big data. *IEEE Systems Journal*, 11(1):88–95, 2015.

[4] Patel, Y.S., *et al.* Cloud of things assimilation with cyber physical system: A review. *Internet of Things: Enabling Technologies, Security and Social Implications*, 93–110, 2021.

[5] Wilhelm, J., *et al.* Review of digital twin-based interaction in smart manufacturing: Enabling cyber-physical systems for human-machine interaction. *International Journal of Computer Integrated Manufacturing*, 34(10):1031–1048, 2021.

[6] Marwedel, P. *Embedded System Design: Embedded Systems Foundations of Cyber-Physical Systems, and the Internet of Things.* Springer Nature, 2021, Cham, Switzerland.

[7] Chen, H. Applications of cyber-physical system: A literature review. *Journal of Industrial Integration and Management*, 2(03):1750012, 2017.

[8] Cao, K., *et al.* A survey on edge and edge-cloud computing assisted cyber-physical systems. *IEEE Transactions on Industrial Informatics*, 17(11):7806–7819, 2021.

[9] Vinel, A., Lyamin, N., and Isachenkov, P. Modeling of V2V communications for C-ITS safety applications: A CPS perspective. *IEEE Communications Letters*, 22(8):1600–1603, 2018.

[10] Rawat D.B., Bajracharya C. *Vehicular cyber physical systems.* Springer. 10:978-3. https://doi.org/10.1007/978-3-319-44494-9, 2017.

[11] Chattopadhyay, A. and Lam, K.-Y. Security of autonomous vehicle as a cyber-physical system. In *International Symposium on Embedded Computing and System Design (ISED)*, pp. 1–6, 2017.

[12] Guo, P., Dusadeerungsikul, P.O., and Nof, S.Y. Agricultural cyber physical system collaboration for greenhouse stress management. *Computers and Electronics in Agriculture*, 150:439–454, 2018.

[13] Bhale, P., Biswas, S., and Nandi, S. *LIENE: Lifetime Enhancement for 6LoWPAN Network Using Clustering Approach Use Case: Smart*

Agriculture. Innovations for Community Services. I4CS 2021. Communications in Computer and Information Science, vol. 1404. Springer, Cham.

[14] Udutalapally, V., *et al.* Scrop: A novel device for sustainable automatic disease prediction, crop selection, and irrigation in internet-of-agro-things for smart agriculture. *IEEE Sensors Journal*, 21(16):17525–17538, 2020.

[15] Zhang, Y., *et al.* Health-CPS: Healthcare cyber-physical system assisted by cloud and big data. *IEEE Systems Journal*, 11(1):88–95, 2015.

[16] Sakr, S. and Elgammal, A. Towards a comprehensive data analytics framework for smart healthcare services. *Big Data Research*, 4:44–58, 2016.

[17] Wu, F.-J., Kao, Y.-F., and Tseng, Y.-C. From wireless sensor networks towards cyber physical systems. *Pervasive and Mobile Computing*, 7(4):397–413, 2011.

[18] Ahmadi, A., Cherifi, C., Cheutet, V., and Ouzrout, Y. A review of CPS 5 components architecture for manufacturing based on standards. In *International Conference on Software, Knowledge, Information Management and Applications (SKIMA)*, pp. 1–6, 2017.

[19] Azarian, M., *et al.* An introduction of the role of virtual technologies and digital twin in industry 4.0. In *International Workshop of Advanced Manufacturing and Automation*. Springer, Singapore, 2019.

[20] Bagheri, B., *et al.* Cyber-physical systems architecture for self-aware machines in industry 4.0 environment. *IFAC-PapersOnLine*, 48(3):1622–1627, 2015.

[21] Bhale, P., *et al.* Brain: Buffer reservation attack prevention using legitimacy score in 6lowpan network. In *International Conference on Innovations for Community Services*. Springer, Cham, 2020.

[22] Arıs, A., Oktug, S.F., Voigt, T. Security of Internet of Things for a Reliable Internet of Services. In: Ganchev, I., van der Mei, R., van den Berg, H. (eds.) *Autonomous Control for a Reliable Internet of Services. Lecture Notes in Computer Science*(), vol. 10768. Springer, Cham. https://doi.org/10.1007/978-3-319-90415-3_13, 2018.

[23] Atlam, H.F., Walters, R.J., and Wills, G.B. Intelligence of things: Opportunities & challenges. In *3rd Cloudification of the Internet of Things (CIoT)*, pp. 1–6, 2018.

[24] Sadiku, M., *et al.* Cyber-physical systems: A literature review. *European Scientific Journal*, 13(36):52–58, 2017.

[25] Madureira, A., *et al.* Empirical validation of Metcalfe's law: How Internet usage patterns have changed over time. *Information Economics and Policy*, 25(4):246–256, 2013.

[26] Bhale, P., Dey, S., Biswas, S., and Nandi, S. *Energy Efficient Approach to Detect Sinkhole Attack Using Roving IDS in 6LoWPAN Network*. Innovations for Community Services. I4CS 2020. Communications in Computer and Information Science, vol. 1139. Springer, Cham, 2020.

[27] Basir, R., *et al.* Fog computing enabling industrial internet of things: State-of-the-art and research challenges. *Sensors*, 19(21):4807, 2019.

[28] Das, A., Dash, P.K., and Mishra, B.K. An intelligent parking system in smart cities using IoT. *Exploring the convergence of big data and the Internet of Things*, pp. 155–180. IGI Global, 2018.

[29] Li, K., *et al.* Internet-based intelligent & sustainable manufacturing: Developments and challenges. *The International Journal of Advanced Manufacturing Technology*, 108(5):1767–1791, 2020.

[30] Alguliyev, R., Imamverdiyev, Y., and Sukhostat, L. Cyber-physical systems and their security issues. *Computers in Industry*, 100:212–223, 2018.

Part 1
Foundations

Chapter 2

Network and Communication Protocols in Cyber-Physical Systems

Hong Liu

Professor of Electrical & Computer Engineering,
University of Massachusetts Dartmouth IEEE Senior Member

hliu@umassd.edu

Abstract: This chapter presents an overview of how CPS components and their users communicate by following an established set of rules referred to as *protocols*. First, we argue the need to network CPS entities and describe the modern network protocols. We divide the protocols supporting CPSs into three categories: machine-to-machine (M2M) connectivity protocols for connecting devices stationary or mobile, routing protocols to find traffic routes between sources and destinations wired or wireless, and M2M communication protocols for specific CPS/IoT applications. Finally, we close the chapter with a CPS example for driverless cars.

Keywords: Internet Protocol Suite, CPS Protocol Suite, connectivity protocols, routing protocols, communication protocols, Wireless Access in Vehicular Environments (WAVE).

1. CPS Protocol Suite and Software-Defined Networking

Entities in a CPS communicate with each other. If imaging each pair required a duplex link for two-way talk, how many links would a community need? As illustrated in Figure 1, the number of links grows

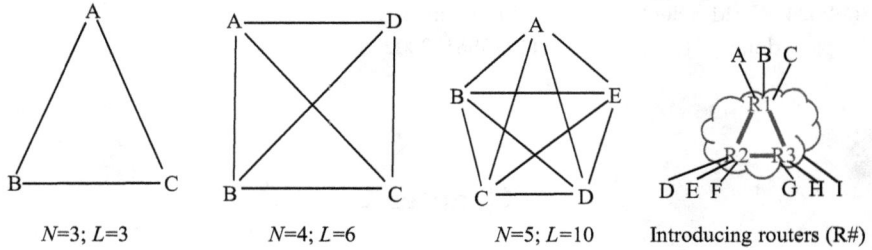

$N=3; L=3$ $N=4; L=6$ $N=5; L=10$ Introducing routers (R#)

Figure 1. Need to network.

Table 1. The Internet protocol stack.

Layer #	Layer Name	Examples
5	Application	HyperText Transfer Protocol (HTTP)
		Domain Name System (DNS)
4	Transport	Transmission Control Protocol (TCP)
		User Datagram Protocol (UDP)
3	Network	Internet Protocol (IP)
		Routing protocols: OSPF, BGP
2	Link	Ethernet: IEEE 802.3
		Wi-Fi: IEEE 802.11
1	Physical	100Base-T, 10GBase-T
		OFDM and sleep modes

quadratically as the population of a community increases. A solution to scale is by introducing intermediate nodes like routers that are tasked to relay data among CPS entities like devices/end users or end nodes. As depicted in the last diagram in Figure 1, by forming a network of end nodes with routers instead of pairwise direct connections, the number of links reduced from $L = 9(9 - 1)/2 = 36$ to 12 for $N = 9$.

1.1.   *The Internet protocol stack: TCP/IP suite*

When CPS entities communicate via networks or the Internet (a network of networks), they need to follow common conventions called *protocols* for mutual understanding. Table 1 shows the Internet Protocol (IP) stack, known as TCP/IP suite, containing five layers of various protocols (Kurose & Ross, 2021). Protocol layering provides conceptual advantages with a structured way to discuss functionalities. However, engineers often design CPS components with cross-layer approaches for efficient use of network resources and adaptive feature of specific applications.

Application Layer: Communication/networking applications and their protocols reside at this top layer of the IP stack. For example, a user requests a web page via the Google Chrome browser to the Amazon web server that replies the document back to the user's browser, following HyperText Transfer Protocol (HTTP).

Another example of application layer protocols is Domain Name System (DNS) that translates user-friendly names of Internet end nodes such as <www.amazon.com> to network/IP addresses like 54.192.189.0. A commercial registrar handles the request for a domain name to verify the uniqueness of the domain name and the availability of the corresponding IP address. The Internet Corporation for Assigned Names and Numbers (ICANN) <www.icann.org> accredits the registrars.

With the IP stack, programmers can easily invent their own application layer protocols for any specific purpose. An application layer protocol runs on multiple distributed end nodes: one end node exchanges packets of information with another end node, using the protocol. A packet at the application layer is called a "message".

Transport Layer: A transport layer protocol transfers application layer messages between *endpoints*, i.e., processes (programs in execution) on end nodes. A packet at the transport layer is called a "segment". The IP stack provides two transport protocols for an application layer protocol to choose: Transmission Control Protocol (TCP) and User Datagram Protocol (UDP).

TCP, chosen by HTTP, for instance, offers a connection-oriented service to its applications. The service guarantees delivery of application layer messages without loss or error. It also contains flow control to match the processing speed of a source node with that of a destination node and congestion control to throttle transmission rates of source nodes from sapping the network.

UDP, chosen by DNS, for instance, offers a connectionless service to its applications. The service is plain vanilla, with no reliability guarantee, no flow control, and no congestion control. The choice of UDP for its transport layer protocol makes DNS resilient against denial-of-service (DoS) attacks. On 21 October 2002, a Distributed Denial-of-Service (DDoS) attack targeted the 13 root DNS name servers, knocking down 7 and severely degrading 2 (Vixie, Sneeringer & Schleifer, 2002). Thanks to the choice of UDP for DNS transport and the wide adoption of caching, most people did not feel the impact in browsing the Internet.

Network Layer: Like the postal service, the network layer moves packets from a source node to a destination node, passing intermediate nodes like routers and gateways. A packet at the network layer is called a "datagram". The IP defines the fields in the datagrams, which dictate how end nodes and routers/gateways act. Each of end and intermediate nodes is assigned an *IP address*, an interface between a node and the Internet like your home address in geolocation.

An IPv4 (IP version 4) address is 4 bytes long expressed in dotted-decimal notation, e.g., 134.88.0.255. In response to the exhaustion of the 32-bit IPv4 address space (2^{32}, ~4 billion different addresses), IPv6 (IP version 6) increases the IP address to 128 bits. The Internet Assigned Numbers Authority (IANA) <www.iana.org> manages IP address space by allocating pools of IP addresses to regional registries. Like your postal address, an interface's IP address has a hierarchical structure (i.e., a network part and a host part, filtered with a subnet mask) and changes whenever you move.

The network layer also contains many routing protocols that find routes to move datagrams from source IP addresses to destination IP addresses. Within an organization's network, called intra-autonomous system (intra-AS), Open Shortest Path First (OSPF) is a widely used routing protocol. All Internet Service Providers (ISPs) use Border Gateway Protocol (BGP) for routing among inter-autonomous system (inter-AS). The service at the network layer is connectionless as the best-effort service model, by which the Internet does its best to deliver packets with no guarantee.

Link Layer: At each hop along a route, the link layer specifies how a sender (end node or router or switch) delivers packets to a receiver (end node or router or switch). A packet at the link layer is called a "frame". The service at the link layer depends on the specific link layer protocol deployed over the particular link — wired or wireless. Note that a local area network (LAN) is considered as a link — wired or wireless — due to lack of routing functionality, despite its name containing "network".

Ethernet is a link layer protocol still dominating the wired LAN market. Many Ethernet technologies have been standardized by IEEE 802.3 Ethernet Working Group <www.ieee802.org/3/>. Ethernet uses Carrier Sense Multiple Access with Collision Detection (CSMA/CD), one of the random access protocol classes for a broadcast LAN, which resolves collisions with binary exponential backoff. Ethernet keeps evolving, from bus topology to hub-based star topology to switched (among point-to-point links, therefore collision-free) for IEEE 802.3z (Gigabit Ethernet).

Another example is Wi-Fi, pervasive wireless LANs as parts of IEEE 802.11 standards by Working Group for Wireless Local Area Networks (WLAN) Standards <www.ieee802.org/11/>. To combat the high bit error rate of wireless channels, 802.11 medium access control (MAC) protocol uses CSMA with Collision Avoidance (CSMA/CA), instead of Ethernet's Collision Detection, and a link layer acknowledgment/retransmission (ARQ) scheme.

The upper network layer could receive different services by different link layer protocols along a route from a source to a destination. Each of node and intermediate node is also assigned a link layer address called a *MAC address*, a unique number to identify its adapter (i.e., its network interface card) like your social security number given at birth. A MAC address is 6 bytes long expressed in hexadecimal notation, e.g., 01-23-45-67-89-AB. The Registration Authority of IEEE <standards.ieee.org/products-programs/regauth/> manages the 48-bit MAC address space. A manufacturer purchases a chunk of the address space containing 2^{24} addresses (the first 24 bits fixed) and varies the last 24 bits for each adapter. Like your social security number, an adapter's MAC address has a flat structure and never changes. Your laptop with an Ethernet card has its MAC address fixed and your cellphone with a Wi-Fi interface also has its MAC address fixed, no matter where you go. Address Resolution Protocol (ARP) translates IP addresses to MAC addresses just like you can find out who lives at which dorm. Conversely, Dynamic Host Configuration Protocol (DHCP) obtains an IP address for your laptop's MAC address when you connect your laptop to the school's network like when you ask which dorm you are assigned to when you start your college year.

Physical Layer: The physical layer moves individual "bits" of a frame from a node to its adjacent node, a hop along a route. The protocols at this layer depend on the protocol choice of the link layer above and the medium choice of the transmission link, wired or wireless.

Take Ethernet, for example" you can select many physical layer protocols for IEEE 802.3 (Kurose & Ross, 2021). It started in the 1980s as 10Base2 (transmission speed of **10** Mbps, **base**band signaling, and a coaxial cable segment not exceeding 100 m with a maximum of **2** segments), 10Base5 (a thick and stiff coaxial cable with a maximum of **5** segments), 10Base-T (a **t**wisted-pair copper wire limited to 100 m), and 10Base-F (a **f**iber-optic cable extending the distance up to 2 km). Transmission speed kept increasing such as Fast Ethernet in the mid-1990s, 100Base-T, for example, and Gigabit Ethernet in the early 2000s, 10GBase-T, for example.

With Wi-Fi, you can also select many physical layer protocols for IEEE 802.11 (Kurose & Ross, 2021). Instead of cables to connect devices in a LAN, Wi-Fi (also referred to as IEEE 802.11 Wireless LAN) uses high-frequency radio waves such as 2.4 GHz, 5 GHz, unused TV bands (54–790 MHz), or 900 MHz. Successive generations in the 30-m range are 802.11 b (11 Mbps) and g (54 Mbps). Brands as Wi-Fi, to compete with 4G and 5G cellular networks, in the 70-m range are 802.11 n or Wi-Fi 4 (600 Mbps), ac or Wi-Fi 5 (3.47 Gbps), and ax or Wi-Fi 6 (14 Gbps). 802.11 af (35–560 Mbps) and ah (347 Mbps) cover the longer range of 1 km aimed at the Internet of Things (IoT) and wireless sensor network (WSN). Note that IoT is often referred as Web of Things (WoT) (Mainetti, Mighali & Patrono, 2015) and (Sciullo *et al.*, 2022).

On the other hand, cellular networks transitioned from circuit-switched mobile telephone network for voice service by 2G to both circuit-switched voice service and packet-switched data service by 3G, and to all-IP packet-switched data network with voice as one of applications by 4G (Kurose & Ross, 2021). As a counterpart of Request for Comments (RFC) by the Internet <www.ietf.org/standards/rfcs>, 3GPP standards organization <www.3gpp.org> developed Long-Term Evolution (LTE) protocol stacks, including 4G LTE and 5G NR (New Radio). Neither 3GPP nor other industry groups have released any standard for 6G at the time of writing this. However, a global race to 6G has already begun, with many countries skipping 5G. LTE uses orthogonal frequency-division multiplexing (OFDM) that combines frequency-division multiplexing (FDM) and time division multiplexing (TDM). LTE also provides additional functions for mobility management and power consumption. An LTE mobile device first attaches to the network through three phases: a bootstrap process for attachment to a base station, a mutual authentication for both the network and the device to know each other's legitimacy, and a data path configuration for the mobile device to send/receive IP datagrams via the base station to/from the Internet through the gateway tunnels. To save power, a wireless device not transmitting or receiving enters a sleep state. LTE mobile devices sleep with two different modes: light sleep in discontinuous reception after several hundred milliseconds of inactivity and deep sleep in idle when inactive for a longer period of five to ten seconds.

1.2. *Encapsulation and decapsulation*

As shown in Figure 2, data flow down the five-layer protocol stack of a source node (e.g., laptop), up and down the three-layer protocol stack of an intermediate node (e.g., router), and up the five-layer protocol stack of a destination node (e.g., webserver). This data path involves a process of *encapsulation* and *decapsulation*.

The laptop's application layer encapsulates a user's text by appending an application layer header information (Ha), which will direct the webserver's application layer to interpret the message when

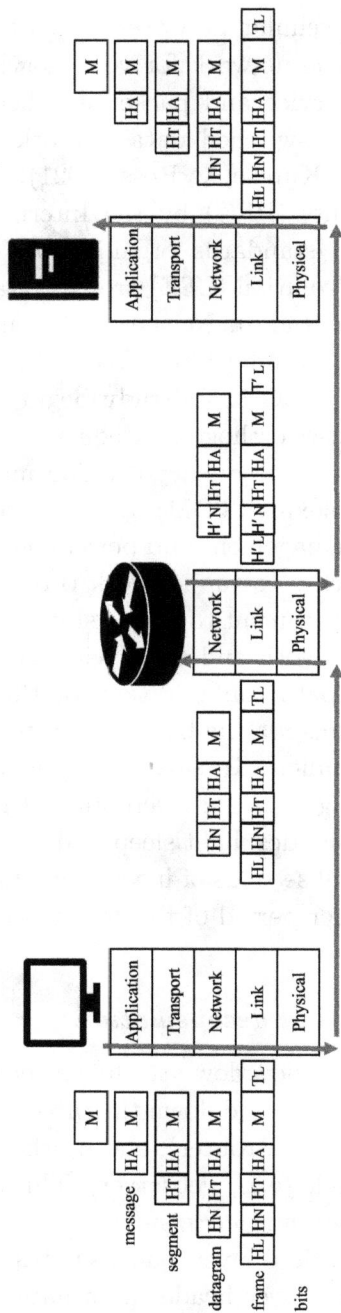

Figure 2. Data flow over nodes through layers.

decapsulated and will pass the message to the laptop's transport layer. The laptop's transport layer encapsulates its application layer's message as a payload by adding a transport layer header (Ht), which will tell the webserver's transport layer in which process by the destination port number to handle the segment. The segment is then passed as a payload to the laptop's network layer that adds a network layer header (Hn). The Hn will inform the webserver's network layer where to reply the webpage requested by the source IP address or the router's network layer where to deliver the datagram by the destination IP address.

Notice that the router starts to participate at the network layer by decapsulating the datagram to obtain the destination IP address for routing and then encapsulating the datagram again with some updates, such as adding a timestamp in the new Hn. The datagram is then passed as a payload to the laptop's link layer that frames a header (Hl) and a tail (Tl). Due to possible change of link protocols along the route, the Hl and Tl of the first hop between the laptop and the router could differ from that of the second hop between the router and the webserver. The frame is finally passed to the laptop's physical layer. Only at the laptop's physical layer will the packet leave the laptop as bits are transmitted over the physical medium.

Figure 3 shows the information captured by Wireshark, a packet analyzer. A user browses a webpage by typing a text:

www.umassd.edu/directory/hliu.html

The laptop's application layer adds Ha with HTTP specification. The transport layer adds Ht with TCP specification. The network layer adds Hn with IP specification. The link layer adds Hl and Tl with IEEE 802.11 specification.

1.3. *CPS Protocol Suite: Cross-layer design*

However, instead of a strict layering architecture used in the World Wide Web, CPS and IoT use cross-layer design due to resource constraints and performance requirements (Jawhar *et al.*, 2017).

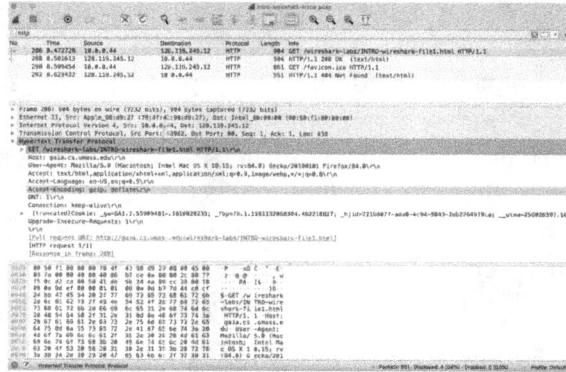

```
http://gaia.cs.umass.edu/wireshark-labs/INTRO-wireshark-file1.html
HA: GET /wireshark-labs/INTRO-wireshark-file1.html HTTP/1.1\r\n HOST: gaia.cs.umass.edu\r\n
HT: Src Port: 53962        Dst Port: 80
HN: Src: 10.0.0.44         Dst: 128.119.245.12
HL: Src: 78:4f:43:98:d9:27 Dst: 00:50:f1:80:00:00
```

Figure 3. Encapsulation of a web GET message.

Table 2. CPS Protocol Suite by functionalities.

Function #	Function Name	Examples
3	Communication	HTTP MQTT
2	Routing	BGP ROLL
1	Connectivity	Wired: IEEE 802.3 Wireless: IEEE 802.11 Mobile: IEEE 802.11p

Cross-layer design allows protocols at different layers to share information and control to achieve efficient use of network resources and adaptive performance for a particular application (Liu & Zhang, 2019). Therefore, this chapter proposes a *CPS Protocol Suite* by functionalities as shown in Table 2: *Connectivity* protocols for machine-to-machine (M2M) connection, *Routing* protocols to find routes between sources and destinations, and M2M *Communication* protocols for specific CPS/IoT applications.

1.4.   *Software-defined networking: Controlling quality-of-service*

As mentioned before, the Internet service model at the network layer is connectionless by "best effort". Datagrams at the network layer could be corrupted, lost, or delayed. Value-added services such as TCP at the transport layer resolve corruption and loss by retransmitting datagrams, which unfortunately causes longer delay. To achieve high quality of service (QoS), intermediate systems such as routers must commit their resources. *Software-defined networking (SDN)* explicitly separates the network layer into two planes: *Control Plane* running protocols to update the routing tables in routers and *Data Plane* forwarding datagrams by each router based on its individual routing table. Figure 4 illustrates SDN's logically centralized control. A remote controller runs a routing protocol to compute routing tables for all routers under its supervision and then distributes the tables to be used by each router. Each router has a local control agent that interacts with the remote controller to configure and manage the router's flow table.

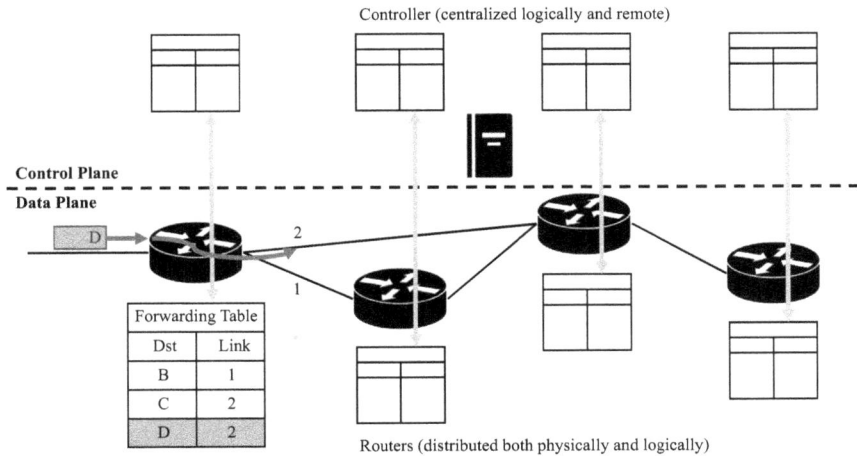

Figure 4.   SDN: Logically centralized control.

Instead of destination forwarding in the traditional Internet, SDN uses generalized forwarding based on OpenFlow standard. A routing table takes the notion of match-plus-action forwarding. The "match" looks up multiple header fields besides destination IP address. In addition to forwarding the packet to the output port(s), the "action" includes filtering packets, translating addresses, and balancing loads. Each entry in the match-plus-action forwarding table is a flow table in OpenFlow terminology. A flow table has a set of header fields to match an incoming packet, a set of counters for bookkeeping, and a set of actions to be taken when matched.

For fault tolerance and performance scalability, the remote controller is conceptually centralized but is implemented with multiple servers (Ma *et al.*, 2019). Figure 5 depicts the components of an SDN controller with its northbound Application Programming Interface (API) for network control apps and its southbound API for routers under its supervision.

SDN is widely adopted by data centers (Google's B4 and Microsoft Research's SWAN), Internet service providers (ComCast's

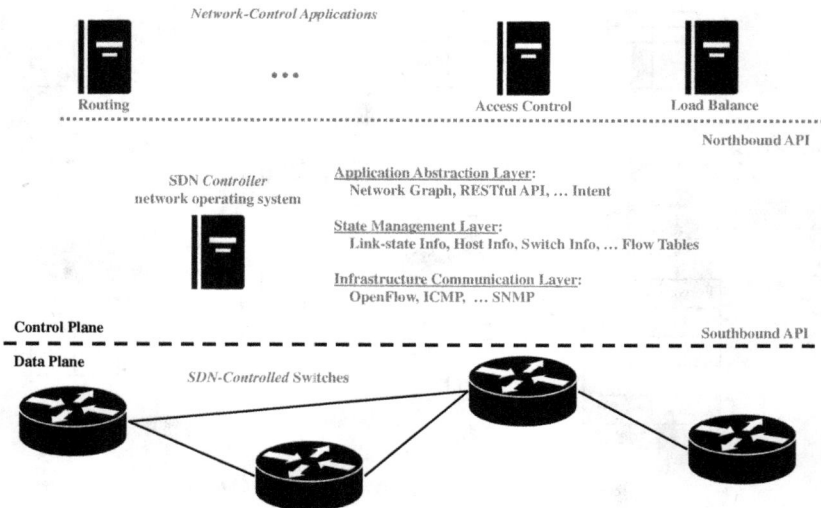

Figure 5.   Components of an SDN controller.

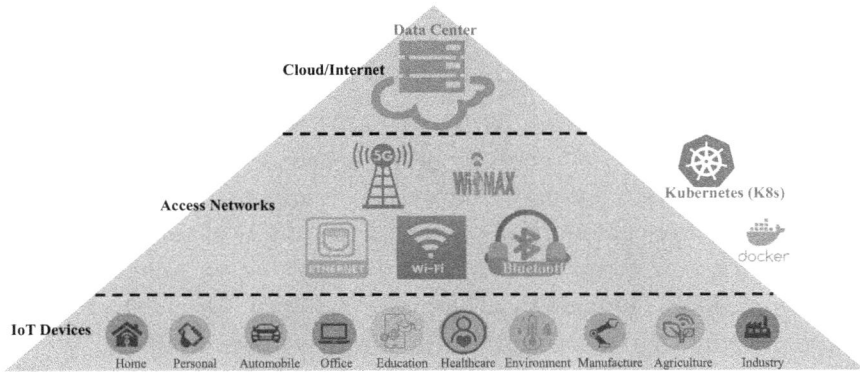

Figure 6.   Edge networking.

ActiveCore and Deutsche's Access), and 4G/5G cellular networks (AT&T and China Telecom).

Access networks physically connect end nodes to the first routers, called "edge routers" of the Internet, on the way to communications among the end nodes. Figure 6 illustrates various types of access networks: data center network, content provider network, enterprise network, café network, home network, LAN, WLAN, mobile WLAN, controller area network (CAN), personal area network (PAN), and cellular network.

To ensure QoS and data privacy, the current distributed computing paradigm shifts computation and storage as close to the point of data consumption. *Edge networking* physically moves computation resources away from data centers or cloud computing toward the edge of the Internet.

2.   Connectivity Protocols

Connectivity in CPS is essential as embedded devices interact with the real-world elements. CPS's entities communicate with each other in a variety of ways: wired or wireless, constrained or mobile, continuous or intermittent (Rajkumar *et al.*, 2010). The *M2M Connectivity Functionality* of the CPS Protocol Suite defines the message format, the exchange order, and the subsequent actions to connect entities of a CPS. Connectivity mostly implements the functions at the Link Layer of the IP stack. However, as mentioned earlier, a CPS protocol models cross-layered design required for both high QoS and strict security under limited computing and networking resources. This section reviews the standards commonly used for connecting machines in three categories: *wired* in need of physical medium such as a twisted-pair copper wire, *wireless* using electromagnetic waves such as radio as opposed to cables, and *mobility* not constrained by space. The latter two categories also address power issues.

Parameters guiding the selection of a connectivity protocol include data rate, distance range, spectrum, topology, Internet support, node support, power consumption, and cost.

Data rate defines the amount of data per unit time a protocol can transfer. In wireless, actual data rates are often much lower than the theoretical maximum due to different factors such as environment in use. For CPS, data rates are classified into levels of *low* (below 1 Mbps), *medium* (about single-digit Mbps), and *high* (above 50 Mbps). Differing from social media applications like YouTube, embedded devices in a CPS such as smart grid do not need high transmission speed but do require periodic or real-time data transfer.

Distance range limits the length a protocol can connect. The coverage extends broadly as PAN, CAN, LAN, metropolitan area network (MAN), and wide area network (WAN). *PAN* connects devices in an individual person's space less than 10 m. *CAN*, developed by the automotive industry, is a bus shared by electronic units on a single vehicle. *LAN* connects devices in a defined area, such as office or home, spreading tens or hundreds of meters. *MAN* covers a city, from 500 m to tens of kilometers. Examples of MAN include a university network or a municipal network. *WAN* goes without limit but is often separated by geographical regions or political boundaries like countries.

Spectrum defines a band of electromagnetic waves for wireless or cellular communications, divided into two categories: free or commercial. *Commercial spectrum* requires license to operate, *free spectrum* does not. The most common free spectrum is *ISM radio bands* — portions of the radio frequency reserved internationally for industrial, scientific, and medical (ISM) purposes. Check regional restrictions even for free bands! For example, a 2.4-GHz ISM band used for Wi-Fi is available worldwide, but a 6.765-MHz ISM band often used for implants is subject to local acceptance. Global System for Mobile communication (GSM), the European telecommunications standard, uses commercial frequency bands including 900 MHz and 1,800 MHz.

Topology indicates ways in which a protocol connects nodes together. Common topologies include bus (tree), star (cluster), ring, mesh, and fully connected. *Bus topology* deploys a central link, called bus, which all nodes attach to. The link is half-duplex that alternates two-way traffics. Each node receives all traffic on the bus and has equal priority to send. *Tree topology* extends from multiple buses formed into a structure of hierarchical tiers. *Star topology* has a central hub which all nodes connect to. The central hub can be designed as simply as a conduit to convey traffic like a logical bus or as complexly as an arbitrator to direct traffic. *Cluster topology* is formed by multiple stars. *Ring topology* connects each node to exactly two other nodes, one on either side to form a one-way loop. *Mesh topology* connects nodes with point-to-point links, forming a connected graph so that any pair of nodes can communicate to each other either directly via their point-to-point link or hopping through links. *Fully connected topology* is actually a full mesh, where each pair of nodes is directly connected via a point-to-point link. Note that hexagon is not a topology but is often used in planning and analysis of cellular networks, because hexagon's tessellating cell shape approximates coverage better than circles.

The Internet support is a necessity for most CPSs, especially when humans are involved in the loop. The Internet's TCP/IP suite provides easy connection with the existing Internet. HTTP, for example, is the protocol for browsing user interface (BUI) in place of graphic user interface (GUI). Radio Frequency Identification (RFID) tags, mainly used for tracking individual items, do not come with Internet support. An extra server can be deployed for RFID tags to connect with, which acts as a gateway to the Internet. Besides connectivity to the Internet, the server can perform additional functions such as security service.

Node support includes both hardware and software. Nodal hardware selects the computing processor, memory storage, and input/output peripherals. Practical issues also need to be considered. For example, although an embedded system has no theoretical limit on the number of ports to connect peripherals, increasing the number could decrease the performance to the point of crashing the system. Nodal software mainly selects the *operating system (OS)* that manages the resources at the node and provides the interfaces of the hardware components with applications. CPS OS tends to be compact designed for a specific purpose, excluding general-purpose OS functionalities such as virtual memory management. Micro-Controller OS (μC/OS) is a real-time OS for microprocessors used in embedded systems (Hill & Culler, 2002). It deploys a preemptive scheduler by priority to satisfy predefined time constraints. TinyOS and OpenWSN, both initially developed by UC-Berkley, are for WSN. Instead of multithreading, TinyOS uses an event-driven programming model where event handlers respond to incoming data packet or sensor reading with run-to-completion semantics (Levis *et al.*, 2005). On the other hand, OpenWSN implements a fully standard-based protocol stack for WSN rooted at IEEE 802.15.4e, a deterministic link layer protocol (Watteyne *et al.*, 2012). RIOT implements a microkernel architecture for low-power wireless IoT devices (Kovatsch *et al.*, 2020). It supports standard programming in C/C++ and common IoT protocols including RPL, 6LoWPAN, and the Internet's TCP/IP suite. However, its multithreading and real-time features might become overkill for WSN (Zimmerling *et al.*, 2017).

Power consumption affects battery life and data throughput of a connectivity protocol. Generally, the lower the power consumption, the lower the data rate. A higher rate makes a device active for longer and the processing faster, hence consuming higher energy. Power consumptions are divided into two categories, idle and active. *Idle* power consumption is a device's rest state, when no communication takes place. *Active* power consumption is a device's state when it is under full communication load, severely draining off the device's battery life. Furthermore, a connectivity protocol provides different power profiles to save battery life: idle state enabling profile, low data rate profile, receive only profile, and transmit only profile, to name a few.

Cost is not a factor to be ignored when selecting a connectivity protocol for CPS or WSN where thousands of devices, even a few cents each, are deployed. Price could become a decisive factor for systems or networks with a large number of nodes.

The rest of the section describes some common Connectivity Protocols with the characteristics defined above. We divide these example protocols in the categories of wired, wireless, and mobile standards.

2.1. *Wired connectivity standards*

IEEE 802 is a family of IEEE standards for computer networks, including LAN, MAN, and PAN, to connect devices. The family contains members, numbered from 802.1 to 802.12 and going on currently to 802.24 (with no significance in ordering but simply the chronic sequence when the standardization project was introduced). Each member has a working group of the IEEE 802 LAN/MAN Standards Committee (LMSC) to maintain. IEEE 802 specifies services and protocols mapped to the lower two layers, Link and Physical, of the IP stack shown in Table 1 of Section 1. It further divides the link layer into two sub-layers: logical link control (LLC) and MAC. For example, at the time of writing this book, the working groups of *IEEE 802.1 Higher Layer LAN Protocols Working Group* and *IEEE 802.3 Ethernet* are active while those working on *IEEE 802.2 LLC* and *IEEE 802.5 Token Ring MAC Layer* were disbanded (Figure 7).

IEEE 802.3 defines the specification of wired *Ethernet*, the most popular family of wired LAN technologies in the market

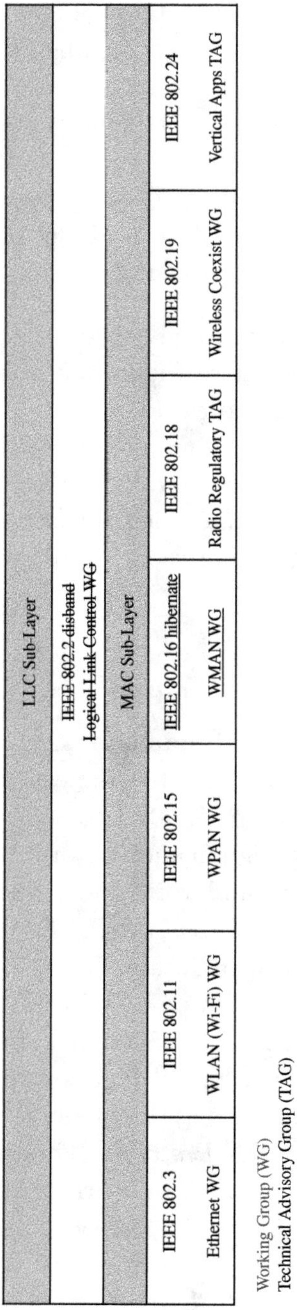

H. Liu

IEEE 802.3	IEEE 802.11	IEEE 802.15	IEEE 802.16 hibernate	IEEE 802.18	IEEE 802.19	IEEE 802.24
Ethernet WG	WLAN (Wi-Fi) WG	WPAN WG	WMAN WG	Radio Regulatory TAG	Wireless Coexist WG	Vertical Apps TAG

LLC Sub-Layer

IEEE 802.2 disband
Logical Link Control WG

MAC Sub-Layer

Working Group (WG)
Technical Advisory Group (TAG)

Figure 7. IEEE 802 family for LAN/MAN/PAN.

with MAN/WAM applications. Embedded development boards, even those at low cost, including 100-Mbps or 2-Gbps Ethernet cards, owe the cheap price to Ethernet. Long hauls often deploy Ethernet for data rates up to 100 Gbps with coaxial cables or 400 Gbps with fiber optics. The original Ethernet LAN was invented in the mid-1970s when power consumption was not a concern. Therefore, Ethernet modules consume a lot of power compared to the other protocols.

A type of connectivity found in telecommunications is *isochronous signal networks* that transmit data in a steady stream of bytes at regular time intervals. Synchronous Optical Network (SONET), standardized by ANSI (American National Standards Institute) and similar to Synchronous Digital Hierarchy (SDH) used in Europe and Japan, encodes electrical signals onto laser light for long-distance propagation over fiber optic WAN. *Asynchronous Transfer Mode (ATM)*, an isochronous signal network, dominated in the 1990s to access the SONET/SDH backbone of public switched telephone network (PSTN) and integrated services digital network (ISDN). Broadband ISDNs carry both high-throughput data traffic (the Internet applications) and real-time/low-jitter multimedia traffic like audio (phone calls) and video (cable TVs). ATM realizes this integrated functionality using asynchronous time division multiplexing. Asynchronous makes empty slots in time division disappear on the output line of multiplexer for efficiency. Unlike computer networks of packet switching to transmit data in packets of various sizes, telecommunications and cellular networks of circuit switching transmit data in packets of a short uniform size called *cells*. ATM uses cells of a small fixed length (53 Bytes), instead of variable-sized frames like Ethernet, which was discussed earlier as a basic transfer unit at link layer. Fixed length enables simple switching and the small size offers flexibility in processing multimedia traffic. ATM also uses a connection-oriented model to set up a virtual circuit, i.e., a soft-state path over packet switches between two endpoints, before exchanging data. Virtual circuit, a combination of circuit switching and packet switching, takes both features for QoS provision and resource sharing by Resource Reservation Protocol (RSVP) to control admissions.

There was an anecdote of how ATM settled in its cell size of 53 Bytes, uncommon in computing and communicating fields as the number is not a power of two. In the late 1980s at CCITT, now known as **ITU**-T, or Telecommunication Standardization Sector of International Telecommunications Union, a conflict arose during the standardization process as to the payload size within an ATM cell: 64 Byte is optimal for American networks while 32 Byte is optimal for Europeans and Japanese networks. CCITT compromised with their average: $(64 + 32)/2 = 48$. Hence, 48-Byte payload plus 5-Byte header totals 53 Bytes total for a cell.

ATM, along with Internet QoS architecture models (IntServ and DiffServ), multimedia apps (Voice over IP and Video on Demand), and other communication techniques, contributed to the dot-com bubble when Internet-based start-ups mushroomed in the mid-1990s. A telecommunication standard to transmit both data and voice, Integrated Services Digital Network (ISDN), became a comic acronym "I Simply Don't kNow" (Figure 8). The dot-com bubble burst in early 2000. Subsequently, ATM was superseded by Next-Generation Network (NGN). Worse yet, before ATM could gain a foothold in wireless and mobile communications, GSM and Wireless LANs (to be discussed in the next subsection) occupied the market.

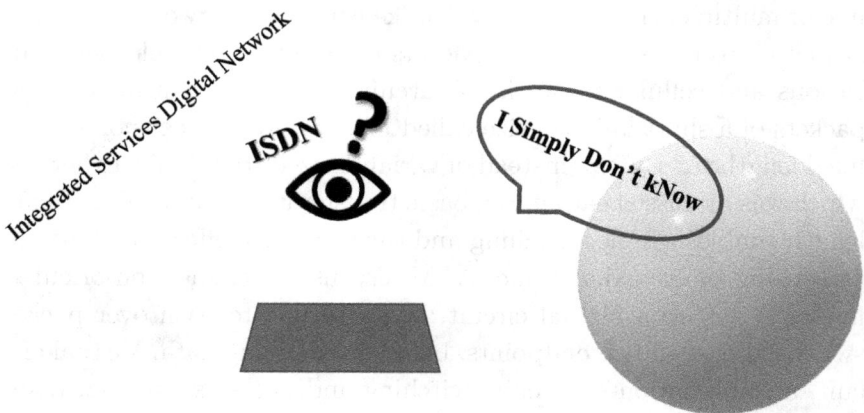

Figure 8. Lessons from the Past: Predict the future.

2.2. Wireless connectivity standards

IEEE 802.11 defines the specification of *Wireless LAN (WLAN)* branded as *Wi-Fi*, the most widely used wireless connectivity standard in the world. The 802.11 family overs MAC of Link Layer and Physical Layer, typically used in conjunction with IEEE 802.2 LLC of Link Layer and designed to interconnect with Ethernet for carrying the IP traffic. The distance spans within 100 m until later protocols increase it to 1 km. Data rate varies from 50 Mbps to 1 Gbps, depending on spectrum and technology. Wi-Fi mainly operates in the 2.4-GHz ISM band but also uses other frequencies such as 5 GHz, 6 GHz, and 60 GHz. Power consumption varies, based on module and manufacturer. In general, Wi-Fi is not power efficient. The number of devices connected to a single gateway (called Wi-Fi hotspot) also varies, restrained by bandwidth, because the system's processor and the connected devices share the bandwidth. A home Wi-Fi device can connect up to 253 devices. The first widely accepted Wi-Fi was 802.11b with range of a 30 m with the maximum data rate of 11 Mbps in 2.4 GHz. Due to the choice of frequency band in 2.4 GHz, 802.11b/g/n/ax devices may occasionally interfere with microwave ovens, cordless phones, and Bluetooth devices. Both 802.11af in unused TV bands (54–790 MHz) and 802.11ah in 900 MHz are targeted for IoT such as sensor networks and meter apps. They offer longer range and wiser power consumption with the downside of lower bandwidth (Figure 9).

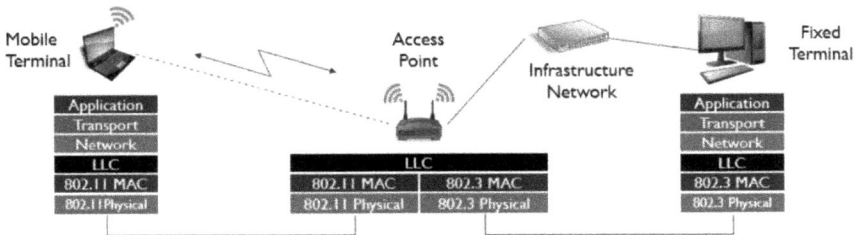

Figure 9. IEEE 802.11.

IEEE 802.15 defines the specification of *Wireless PAN (WPAN)*, connecting wirelessly with fixed, portable, and moving devices within the personal operating space (tens of meters or less) at low power and low cost. Its first task, *IEEE 802.15.1*, is based on Bluetooth technology that defines MAC of the Link Layer and Physical Layer of the IP stack. The Bluetooth Special Interest Group (Bluetooth SIG), a global non-profit organization of more than 36 thousand companies, manages the development of Bluetooth standards and its licensing. *Bluetooth* is a short-range low-bandwidth WPAN protocol. It operates in the unlicensed 2.4-GHz ISM band and uses frequency-hopping spread spectrum. Older Bluetooth standards offer data rates as low as 720 Kbps for Bluetooth 1 and an Enhanced Data Rate (EDR) of 3 Mbps for Bluetooth 2. Newer standards increase data rates to 34 Mbps for Bluetooth 4, called *Bluetooth Smart* that includes *Classic Bluetooth* of legacy Bluetooth protocols, *Bluetooth High Speed* based on Wi-Fi, and *Bluetooth Low Energy (BLE, previously known as Wibree)* for very low power consumption. Despite a similar name, BLE is a completely different protocol with a data rate of only 1 Mbps to send periodic data packets. BLE is very power efficient: a BLE device can perform over million transactions using a coin battery. BLE provides security with up to 128-bit AES encryption. It also imposes no theoretical limits on the number of nodes connected to an access point. However, BLE is not suitable for voice calls or music streaming. The Bluetooth 5 series mainly focuses on IoT with incremental new features: mesh networking added in Bluetooth 5.1, Low-Energy (LE) Audio in Bluetooth 5.2, and enhanced channel classification and encryption key size control (Wan *et al.*, 2018) in Bluetooth 5.3.

Another popular task of IEEE 802.15 is *IEEE 802.15.4 (Low Rate WPAN)* for low data rate and low protocol complexity with very long battery life (months or years). Among standardized and proprietary protocols run over IEEE 802.15.4 at Network Layer or as Mesh LAN, ZigBee is the most common standard for low-power mesh networking nodes. Although its range is up to 100 m, ZigBee can span connectivity over longer distances by its mesh networking. Unlike other connectivity protocols, the ZigBee protocol stack also defines functionalities at Network Layer and Application Layer. Therefore, ZigBee is not only a connectivity protocol but also a communication protocol. The latter will be discussed in Section 4.

Besides data networks like ZigBee, IEEE 802.15.4 supports various special-purpose connectivity protocols including cellular networks such as 6LoWPAN. 6LoWPAN stands for IPv6 over Low-power Wireless PAN. As the name suggests, it is a standard that enables the use of IPv6 on low-power edge devices by transmitting IPv6 packets over IEEE 802.15.4. Native to the IP stack, 6LoWPAN devices and networks do not require any special gateways to connect to other IP networks. Therefore, 6LoWPAN offers interoperability and simplicity in design compared to ZigBee or Bluetooth. The Internet Engineering Task Force (IETF.org) specifies 6LoWPAN in a series of RFC documents: RFC 4919 describes the problem statement and RFC 4944 defines frame format and delivery (updated by RFC 6282 header compression, RFC 6775 neighbor discovery, and more). 6LoWPAN is ideal for those CPSs where IP connectivity and low power consumption are essential.

Unlike computer networks using one main transmitter at a node, cellular networks deploy multiple small interconnected transmitters or *cells*. Cellular technologies also reuse frequency: one frequency being used by multiple cells to increase range and capacity. However, these cells must be geographically far apart from each other to minimize interference. Cell phone providers call each generation of wireless technology as "G". The first generation or 1G was analog cellular phones. 2G switched to digital for voice and texts in the 1990s. 3G increased data rates in the 2000s. 4G started in 2010 with the LTE high-speed wireless communication standard for data terminals, which then replaced the primary service on 2G and 3G networks for mobile phones with voice over 4G or VoLTE (Voice over LTE). Currently, carriers are installing 5G while 6G is under development.

For every generation, there are multiple competing standards. The two main competitors for 2G and 3G are CDMA (Code Division Multiple Access) and GSM (Global System for Mobiles). The two main competitors for 4G are WiMax (Worldwide Interoperability for Microwave Access) data networks and LTE cellular networks.

Qualcomm, a US-based global semiconductor and telecommunications company, invented CDMA, the best technology in the mid-1990s for switching from 1G analog to 2G digital. GSM was brought out around the same time by an industry consortium. CDMA is mostly limited in the United States while GSM is used worldwide. Both CDMA and GSM use multiple access technology, with which calls share the same medium. GSM uses TDMA or "time division" for calls to take turns while CDMA uses "code division" for calls to talk in different languages. To consumers, the difference between GSM and CDMA phones lies in whether a device uses a removable SIM (subscriber identity module) card or a smart card for storing identification information with a specific mobile network pinpointing a device. Since GSM phones use SIM cards, you can buy a new phone with your phone number and plan intact by taking the SIM card from your old phone. CDMA, on the other hand, used embedded serial numbers of devices to identify their subscribers to carriers. At the inconvenience of users, carriers had more control over choosing phones and accessing their mobile networks. Note that 3G GSM adopts code division technology, called wideband CDMA (WCDMA) or Universal Mobile Telephone System (UMTS).

LTE not only became the globally accepted 4G wireless standard but also ended the CDMA/GSM split. The LTE radio access network uses a combination of FDMA and TDMA on the downstream channel, known as orthogonal frequency-division multiple access (OFDMA). "Orthogonal" means that signals are scheduled to be sent in time slots (as small as 0.5 ms) on different frequency channels to minimize interference, even when channel frequencies are tightly spaced. 4G LTE is very mature now. 5G-NR (New Radio) is coming as a single global standard. In any case, your choice of connectivity technology is based on the range and data rate you need, local availability, or cost.

IEEE 802.16 defines the specification of WirelessMAN Air Interface for Broadband Wireless Access, branded as Worldwide Interoperability for Microwave Access (WiMAX). Competing with LTE for 4G, WiMAX is a standard-based technology for the delivery of last-mile wireless broadband access replacing cables and telephone lines. Like all IEEE 802 series, 802.16 standardizes MAC of Link Layer and Physical Layer of the IP stack. It operates on bandwidths between 1.25 MHz and 20 MHz and supports multiple-input multiple-output (MIMO) antennas for better propagation characteristics and higher data rate. Like the LTE radio access network discussed above, WiMAX uses OFDMA. 802.16 MAC contains several convergence sub-layers, specifying how wired technologies such as Ethernet and ATM encapsulate on the air interface. It also includes secure communications with AES encryption, power-saving management like sleep/idle mode, and handover mechanisms.

One of RFID applications functions as an advanced replacement for barcodes. Instead of a barcode, an RFID tag is placed on a product. Instead of using light for a barcode reader to scan the code, an RFID reader interacts with tags using radio waves. Their wireless communications possess several advantages: longer range (typically 1 m), simultaneous multiple scans (hundreds of tags being read at once), and not requiring line of sight. Tags can be implanted anywhere in many objects for a reader to track as long as they are within the range. A tag contains a tiny memory card (some kilobytes) to store its identifier and other information, a microchip to process requests and communicate data, and antennae to transmit and receive RF signals. Tags are classified as either passive or active. *Passive* tags take their power from the RF signals transmitted by a reader. *Active* tags contain batteries to power them, offering much longer range for a price.

2.3. *Mobile connectivity standards*

All wireless connectivity standards described in the previous subsection offer mobility either limited by range or extended with handover among fixed-network infrastructures. We will discuss handover in Section 3 of routing protocols. This subsection, particularly, covers connectivity to mobile network infrastructures.

IEEE 802.11p, an amendment to IEEE 802.11 WLAN/Wi-Fi for wireless access in vehicular environments (WAVE), was approved in

2010. 802.11p uses the 5.9-GHz band licensed by the U.S. Department of Transportation (USDOT) Intelligent Transportation Systems (ITS) as Dedicated Short-Range Communications (DSRC). The U.S. Federal Communications Commission (FCC) allocated 75 MHz of the DSRC spectrum in 1999, and the European counterpart allocated 30 MHz in 2008. DSRC applications include toll collection and vehicle safety. During the twenty-one years of its existence, about fifteen thousand vehicles in the U.S. (only 0.0057%) were equipped for DSRC. Because the auto industry failed to utilize the spectrum, FCC in 2020 reallocated the lower 45 MHz half of the DSRC spectrum (5.850–5.895 GHz) for Wi-Fi and other unlicensed uses. IEEE 802.11p was superseded and subsequently incorporated into the base standard, IEEE 802.11 WLAN/Wi-Fi.

Because the connectivity among vehicles and/or roadside infrastructure exists for a short time interval, the amendment defines a method to exchange data without the need to establish a priori service set. A service set is a group of devices sharing a service set identifier (SSID), via association and authentication procedures, to form a Wi-Fi network. Using the wildcard basic SSID (a value of all 1s) in the frames' header, high-speed vehicles exchange data frames among them and/or with a roadside unit as soon as they arrive within the proximity. Because vehicles are not authenticated by IEEE 802.11 and its amendment, upper layers must provide such functionality, as defined by *IEEE 1609 Family of Standards for WAVE*, to be presented in Section 5 as a use case.

As a daring research and development project, Google started Project Loon in 2011 to build balloon networks that beam the Internet to un/under-connected rural regions and to disaster-stricken areas. Nine years later in January 2021, Google canceled Project Loon due to difficulty in making a profit (Alexander *et al.*, 2021). This aerial wireless network reached a data rate of 155 Mbps over a distance of 100 km with laser links between balloons and delivered speeds to end users comparable with LTE (Serrano *et al.*, 2021). Loon developed a radical new technology for mobile networking shown in Figure 10. The network data plane connected LTE users with Internet antennae and transferred their data through stratospheric

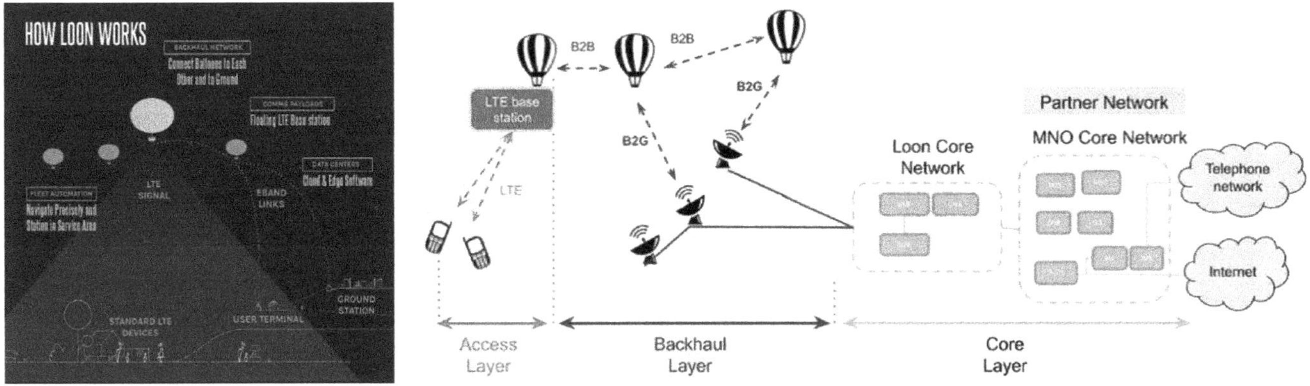

Figure 10. Loon LTE network architecture: Core, Backhaul, and Access layers.

balloons (B2B) to a ground station (B2G) connected to an ISP of the global Internet. The SDN control plane managed service requests, authentication, routing, weather ingestion, and other functionalities.

Like a typical mobile network, the Loon LTE network architecture had three primary network layers shown in Figure 10: **Access Layer** connects the LTE base stations to the user's phone as the last hop. **Backhaul Layer** consists of everything between the base stations and the core network. **Core Layer** is a networking facility including the Mobile Network Operator (MNO) connecting to the partner network such as Telephone network or the Internet. Loon differed from other LTE networks in two key areas: Loon's base stations, instead of fixed on buildings, floated on balloons at high altitude across countries; Loon's backhaul for each base station, instead of one fiber link from the base station to the MNO, used multiple links from Loon's ground stations across the service area (each of which tracks the moving balloons). Loon's backhaul network contained balloon-to-balloon (B2B) laser links, balloon-to-ground (B2G) microwave radio links, ground stations, and fiber links from ground stations to Loon's core network. Each balloon, the size of a tennis court, provided 4G LTE coverage over $11,000 \, km^2$ — an area equivalent to the coverage by 200 cell towers. A balloon self-navigated in the stratosphere about $20 \, km$ (12 mi) high by adjusting its altitude to float in a wind layer at the desired speed and in the desired direction. Note that typical aircrafts fly around $15 \, km$ (9 mi) and satellites orbit above $100 \, km$ (60 mi).

Loon debuted in 2017 at Puerto Rico, providing wireless coverage after Hurricane Maria devastated the island's telecommunications infrastructure. The *ad-hoc* network delivered emergency Internet service. In 2019, Loon launched its first commercial telecom tryout over Kenya with 25 balloons and 6 ground stations for 3 sites. Deployment followed in 2020, providing mobile phone service to the country's rural millions. The floating balloon network saved the heavy infrastructure investments of building the Internet on tough terrain. Although Loon became defunct, the efforts of stratospheric balloon networks are ongoing in government, industry, and academia. To monitor infrastructure health or survey border security, low-altitude drones cover a small area while outer-space satellites take images that lack resolution. Stratospheric balloons, at a fraction of a satellite's price and at an altitude high enough not to collide aircrafts and low enough to observe a vast area in sufficient detail, provide an ideal solution to surveillance. The U.S. military deployed stratospheric balloon networks across Midwest states to detect homeland security threats. NASA applied Loon SDN to its next-generation space communications architecture for interoperability (Barritt & Cerf, 2018). NSF funded New York University to study atmospheric gravity waves using Loon high-altitude balloons and machine learning techniques, improving short-term forecasts of weather extremes and long-term projections of climate changes.

3. Routing Protocols

The IP stack conceals routing functionality from end systems such as users and devices. Its layered architecture benefits networking application programmers by encapsulating lower layers' details and routing at the network layer, so that they can focus on developing applications. CPS protocol suite, on the other hand, adopts cross-layer design for efficiency. Therefore, it needs transparency in selecting all functionalities, routing and connectivity included. Additionally, the nature of technic heterogeneity involves multidisciplinary fields of computation and communication. For example, handover procedures in telephony networks equate to routing functions in computer networks.

Factors of routing in general include ways of traffic flows, routing algorithms, policy on intra- vs inter-autonomous systems or home vs visited networks, and control mechanics on delay and reliability (Bhatia *et al.*, 2021).

Traffic flows can choose among four ways: unicast, multicast, broadcast, and anycast. *Unicast* sends data to a specific node, as one-to-one communication. *Multicast* sends data simultaneously to a specified group of nodes, by a multicast group name. *Broadcast* sends data simultaneously to all nodes, by wildcard matching. *Anycast* sends data without specifying any node, received by the nearest node in a collection of nodes (servers at multiple locations) sharing the same IP address. Here, "nearest" is measured topographically, not geographically, though the two are often the same. Often, anycast and broadcast are implemented with *flooding* when a router sends an incoming packet to each of its links except the one on which the packet came.

Routing algorithms determine least-cost routes from senders to receivers through the network of routers, as a graph of nodes and links. The total cost of hops along a route includes link metrics and node status. Routing algorithms can be classified by multiple dimensions: centralized vs distributed process, static vs dynamic to network changes, load-sensitive or not to cost changes. For example, *link-state algorithms* are *centralized* while *distance-vector algorithms* are *distributed*.

Autonomous systems of the Internet or service providers of cellular networks determine different policies for routing. A domain/institution or a service provider can solely decide all the factors of routing for *intra-AS* or *home networks*. However, multiple domains/institutions or service providers must cooperate and negotiate how to run *inter-AS* or *visited networks*.

Control mechanics for QoS evolve with network architectures. As discussed in Section 1, *SDN* separates control mechanics on delay and reliability between forwarding in data plane and routing in control plane. Other control mechanics deploy a simple tool like *Internet Control Message Protocol (ICMP)* or involve a complex database system like *Simple Network Management Protocol (SNMP)*.

This section reviews the standards commonly used in finding a route from a source machine to a destination machine over multiple hops in networks. Like the previous section on Connectivity Protocols, we also divide Routing Protocols into the categories of wired, wireless, and mobile.

3.1. *Wired routing standards*

TCP/IP protocol suite contains general-purpose routing protocols, supporting various service models over long hauls in wired mesh networks. They are complex and inconsiderate about energy efficiency. Note LAN is not a network but a link, despite its name containing "network". CPS developers usually do not get a chance to select a routing protocol in wired mesh networks for long hauls due to their transparency. Knowing what to connect as intra-AS or inter-AS is sufficient. Intra-AS routing in the Internet widely uses OSPF, a link-state protocol that floods link cost information to establish the network topology and runs Dijkstra algorithm at the logically centralized SDN controller to find the least-cost path. Inter-AS routing uses BGP, a distributed and asynchronous protocol based on distance-vector algorithm.

3.2. *Wireless routing standards*

Unlike wired mesh networks such as the Internet for supporting various service models over long hauls, wireless networks do not need complex routing methods. For wireless networks, a routing process involves three simple steps. First, find neighboring nodes. Instead of node IDs such as IP addresses in the Internet, nodes in wireless networks are characterized by location coordinates, transmission capability, remaining energy, and more. Second, select the best node based on their characteristics. Third, send messages to the chosen node. Obviously, a sleeping node would be awakened before sending it messages. In addition to routing factors such as control mechanics and node characteristics like remaining energy, wireless networks also consider data aggregation to reduce transmission cost.

Routing Over Low power and Lossy (ROLL) network is an IETF working group on routing solutions for Low power and Lossy Networks (LLNs). LLN interconnects embedded devices of limited power, memory, and processing resources with heterogeneous links. Links include IEEE 802.15.4 Low-Rate wireless PAN (e.g., ZigBee, MiWi), IEEE 802.15.1 Bluetooth, IEEE 802.11 wireless LAN and Mesh (e.g., low-power Wi-Fi), or other low-power Power Line Communication (PLC) wires. The work focuses on IPv6 only and an end-to-end solution to ensure interoperable networks. In addition to optimize power and reliability for multihopping, the work also addresses routing security and manageability (e.g., self-configuration) issues.

The ROLL working group developed *RFC 6550 IPv6 Routing Protocol for LLN (RPL)*. LLN routers operate with constraints on processing, memory, and energy (battery power). Their interconnects are characterized by high loss rates, low data rates, and instability. RPL supports unicast (point to point between two devices), multicast (point to multipoint from a central control device to a subset of devices), and reverse-multicast (multipoint to point from devices towards a central control device) for LLN from a few dozen to thousands of routers. As SDN separates a data plane from a control plane, RPL separates packet processing/forwarding from routing optimization. Examples of optimal routing objectives include minimizing energy, minimizing latency, or satisfying constraints. Differing from the traditional Internet, LLN devices/nodes can play the roles of both *host* (generating but not forwarding RPL traffic) and *router* (both forwarding and generating RPL traffic).

RPL discovers links and then selects peers sparingly, because LLN does not predefine topology as in radio networks. Since LLN Layer 2 ranges overlap only partially, RPL forms non-transitive/Non-Broadcast Multi-Access (NBMA) network topology, upon which RPL computes routes. Specifically, the RPL network topology adopts *Destination-Oriented Directed Acyclic Graphs (DODAGs)* by partitioning a Directed Acyclic Graph (DAG) at roots, one DODAG per root. A *DAG root* is a node within the DAG that has no outgoing edge. Because the graph is acyclic, by definition, all DAGs must have at least one DAG root and all paths terminate at a DAG root. A *DODAG root* is the DAG root of a DODAG. The DODAG root may act as a border router for the DODAG; in particular, it may aggregate routes in the DODAG and may redistribute DODAG routes into other routing protocols. Therefore, RPL routes are optimized for traffic to or from DODAG roots that act as sinks for the topology. ICMPv6, RFC 4443 ICMP for the IP Version 6, carries RPL control messages:

- DODAG Information Solicitation (DIS) discovers nearby DODAG;
- DODAG Information Object (DIO) responds to DIS message or periodically refreshes the information of the nodes on the network topology;
- Destination Advertisement Object (DAO) updates the information of parent nodes. A *DODAG parent* of a node within a DODAG is one of the immediate successors of the node on a path toward the DODAG root. A node's *Rank* defines the node's individual position relative to other nodes with respect to a DODAG root. Rank, computed as link metrics and/or node constraints depending on the DAG's Objective Function, strictly increases in the Down direction and strictly decreases in the Up direction. Thus, a parent's Rank is lower than the node's Rank.

RPL has been implemented in WSN for various OSs. The most common OS for its implementation is Contiki, an OS for networked memory-constrained systems of low-power wireless IoT devices (Watteyne *et al.*, 2012). Other OSs include TinyOS and RIOT, discussed in Section 2. Due to its complexity, RPL has not been in wide use.

Researchers have proposed various simple routing approaches for wireless networks, aiming at energy efficiency. *Low-Energy Adaptive Cluster Hierarchy (LEACH)* integrates a MAC protocol of TDMA and CDMA with hierarchical clustering and simple routing. LEACH divides routing operation into rounds. During each round, a cluster head creates a schedule for each node in its cluster to transmit data. All non-cluster nodes only communicate with the cluster head in the scheduled TDMA. They only need to keep their radios on during their assigned time slot, minimizing power consumption. Cluster heads aggregate and compress the data before forwarding to the base station (sink) with different CDMA codes, eliminating interference.

At the end of each round, a node (that has not been a cluster head in a predefined number of rounds) determines if it will become a luster lead for the next round. Each non-cluster node selects the nearest cluster head to join the cluster for the next round of data transmission.

Power-Efficient Gathering in Sensor Information System (PEGASIS) improves LEACH by creating a sensor chain, instead of forming clusters. After a chain is formed at the end of each round, a leader with the most residual energy is chosen from the chain. During each round, only the leader in the chain is in charge of transmitting data to the base station (sink). All other nodes communicate only with their nearest neighbors, determined by adjusting signal power to only hear the nearby nodes. PEGASIS eliminates many shortcomings of LEACH: lack of considering residual energy when selecting cluster heads, uneven distribution of cluster heads, variable size cluster formations, and missing the case when a single hop to base station is more energy efficient than going through cluster head. PEGASIS extends the lifetime of a wireless network to about twice that of LEACH.

Virtual Grid Architecture (VGA) forms a grid topology on top of a wireless network. It then utilizes data aggregation and in-network processing to maximize the network lifetime in terms of energy consumption.

3.3. *Mobile routing standards*

A mobile device changes its access point to a network over time. There is a spectrum of mobility from a network layer perspective, depending on the active degree of a device as it changes its access point. This subsection constrains *mobility* as transferring the association of a moving device with access points among WLANs or LTE cells while maintaining the device's ongoing connection, i.e., the device keeping its active state. Therefore, a device is not considered mobile if it disconnects from a wireless network and powers down before reconnecting to another wireless network. From a network layer perspective, a moving device is also not considered mobile if it remains attached to the same wireless access network. From a link layer perspective, a moving device is not mobile if it keeps its association with the same access point such as a Wi-Fi hotspot or LTE base station.

According to Global System for Mobile Communications (GSM) Association (GSMA), mobility in wireless networks is defined as *handover* among access networks by a single service provider or *roam* across access networks by multiple service providers. It is similar to Internet routing as among intra-AS or across inter-AS in wired mesh networks.

In cellular WANs (4G, 5G, and beyond), each subscriber associates with a *home network* by a service provider, called *Home Subscriber Service (HSS)*. HSS stores the information about each subscriber: a 64-bit International Mobile Subscriber Identity (IMSI) and a phone number in a SIM card, cryptographic keys, service information, accounting, and billing. When a device is connecting to a non-home network, i.e., *roaming to a visited network*, both the home network and the visited network coordinate for the subscriber's communication. This home network association for a mobile device has two advantages. First, the home network provides a single source to find the device's vital information, with no data sharing with other parties to ensure security and privacy. Second, the home network serves as central coordination to communicate with a roaming device, establishing trustworthy data exchanges without violating privacy.

Via a home network, roam enables mobile access to a visited network by a service provider with which the device does not have any prior arrangement. The mobile network architecture adopts two basic approaches, direct and indirect routing. With *indirect routing*, a host or another device corresponds with a mobile device by sending a message to the mobile device's permanent home address along with its IMSI. The home network router of the mobile device intercepts the message, consults the HSS to find out where the mobile device is currently residing (at its home network or with a visited network), and forwards/tunnels the message (in the latter case) to the visited network router which then delivers the message to the mobile device's temporary visited address. With *direct routing*, a host or another device first consults the HSS to find the visited network where the mobile device is currently residing and then directly sends/tunnels the message to the visited network router. Each of the two routing approaches has advantages and challenges. Indirect routing may suffer from inefficiency known as *the triangle routing problem* when both devices are residing at the same visited network: the two wait for the home network to forward/tunnel data to the visited network, instead of exchanging data directly within the visited network. Direct routing adds complexity: requiring protocol mechanisms to locate and update whereabouts of the mobile device for the host.

The Internet itself does not support mobility like cellular networks. However, IETF has standardized IP Mobility since the mid-1990s. RFC 5944 specifies the *Mobile IP* architecture and protocols. Each mobile node has its permanent *home address* regardless of its current access point to the Internet. When moved to a *foreign network* away from its home, the mobile node is associated with a *care-of address* with its current access point to the Internet. A node corresponds with a mobile node by sending datagrams to the mobile node's permanent IP address. A *home agent*, where the care-of address is registered, forwards/tunnels each datagram to the case-of address for the mobile node. However, Mobile IP has been hardly deployed in practice use. It missed the business opportunities due to the lack of use cases, dampened by the flourish of cellular networks.

Fortunately, several organizations carry on other activities for the Internet to authenticate access across IP foreign networks. RFC 3579 defines *Remote Authentication Dial In User Service (RADIUS)* support for the Extensible Authentication Protocol (EAP). RADIUS provides an authentication framework to support multiple authentication mechanisms, in which the Network Access Server (NAS) forwards EAP packets for the RADIUS server to run method-specific authentication code. EAP was originally developed for use with Point-to-Point Protocol (PPP) to connect computers with the Internet or the telephone network, but it is now also in use with the IEEE 802 LAN/MAN family. On the other hand, IEEE 802.1X defines *Port-Based Network Access Control* for LAN/MAN. Port-based network access control allows a network administrator to restrict the use of service access points (ports) to secure communication between authenticated and authorized devices. It supports mutual authentication between the clients of ports attached to the same LAN, including discovery and establishment of the security associations.

One worldwide roaming access service is *eduroam* (education roaming). This Wi-Fi roaming service by eduroam is secure and free of charge to users. Any user from an eduroam participating site gets Internet connectivity across campus and when visiting any other participating institutions. User credentials are kept by the user's home institution only and are not shared with visit sites. The participating National Research and Education Networks (NRENs) securely route requests to the server run by the user's home institute for verification and validation.

The eduroam technology adopts 802.1X to provide an authentication mechanism for devices to connect with a wired or wireless LAN while a linked hierarchy of RADIUS servers maintains user credentials. Participating institutions must operate RADIUS infrastructure and agree to the terms of use. As shown in Figure 11, a user

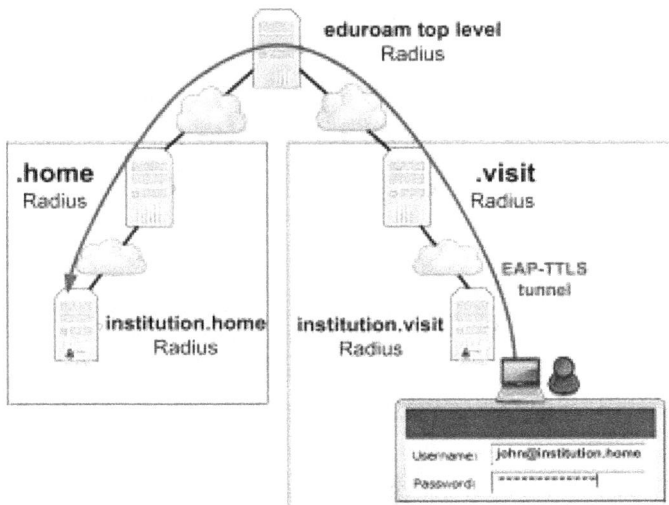

Figure 11. RADIUS supporting eduroam.

at a visit site types in his/her credentials (username and password); the RADIUS hierarchy forwards the user credentials securely to the user's home institution where verification and validation are conducted. Data encryption standards are used to protect the traffic from the user's device over the wireless network. The user's home institution maintains and monitors user information, even when the user visits another institution. Thus, eduroam ensures security and privacy of Wi-Fi roaming service to the Internet.

Handover, instead of roaming, maintains ongoing connections while a user moves across different cellular WANs (4G, 5G, and beyond) by a single service provider. Reasons for handover include the signal from the current base station fading away; the current cell becoming overloaded; and the mobile device reporting better measurements of nearby base stations. GSMA documents refer to the 3rd-Generation Partnership Project (3GPP) for handover specifications. LTE networks provide various types of handover:

- *Intra-LTE handover* when the source/current base station and the target base station are in the same LTE network, i.e., same Mobility Management Entity (MME) and same Serving Gateway (SGW). Its use cases include handover using *X2 interface* or *S1 interface*.
- *Inter-LTE handover* toward other LTE nodes, as *inter-MME* involving two MMEs (source and target) but the same SGW or *inter-MME/SGW* involving two sets of MME/SGW (source and target).
- *Inter-RAT handover* between different radio technologies, e.g., handover from LTE to wideband CDMA.

Figure 12 depicts a simple process of intra-LTE handover:

Step 1. Source base station selects Target base station and sends it a Handover Request message.

Step 2. Target base station pre-allocates the resources, if there are sufficient, for the mobile device and replies a Handover Request Acknowledge message to Source base station.

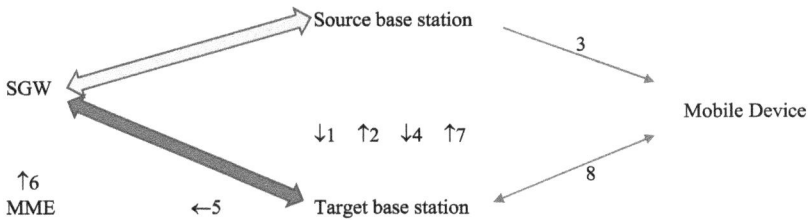

Figure 12. Simple handover process.

Step 3. Source base station, upon receiving the acknowledgment from Target base station, informs the mobile device to communicate with the new base station from now on.

Step 4. Source base station forwards the rest of tunneled datagrams received from SGW to Target base station, instead of to the mobile device.

Step 5. Target base station informs MME that it is the new source for the mobile device.

Step 6. MME relays the handover information to SGW which reconfigures the tunnel to Target base station – the new source – and releases the tunnel to the old Source base station.

Step 7. Target base station confirms the handover completion to the old Source base station which releases the resources allocated for the mobile device.

Step 8. Target base station and the mobile device can begin exchanging datagrams tunneled from/to SGW. Note that the datagrams include those forwarded by the old Source base station to Target base station during handover Step 4.

4. Communication Protocols

The traditional Internet supports information exchanges among human beings for some special purpose, for example, HTTP for browsing webpages and SMTP for transferring emails. Therefore, communication protocols have been specified at the Application Layer of the TCP/IP protocol suite. Likewise, CPS needs various

M2M communication protocols to facilitate the transfer of messages among CPS components for different purposes (Sha *et al.*, 2008). The *M2M Communication Functionality* of the CPS Protocol Suite defines the standards of naming, messaging, and controlling. Naming specifies how each machine or device is to be identified and located. Messaging defines the representation format and the processing convention of messages so that all machines can understand. Controlling enables the participating machines to cooperate among them and to get the network providing QoS and security assurance.

Parameters guiding the selection of an M2M communication protocol include communication mode, implementation mode, transport layer support, payload, service discovery, device management, QoS, security, and privacy.

Communication mode defines a paradigm/pattern that a protocol follows. A common mode is *request/response interaction* model. A *client* requests to a central server for data or services, and the *server* responds in a predefined pattern depending on the protocol. A single request may yield multiple responses, to a multicast request, for example. Another communication mode is *publish/subscribe* messaging, "pub/sub" in short. Pub/sub forms asynchronous service-to-service communication without a server. Components called *publishers* simply push their messages on a topic, to which other components called *subscribers* receive the messages immediately. A sibling of pub/sub is *message queue*. Message queue also forms asynchronous service-to-service communication without a server. A component called *producer* adds a message (usually small) to a queue, from which another component called *consumer* retrieves the message. Although many producers and consumers can use the queue, each message is processed only once by a single consumer. The main difference between pub/sub and message queue lies in whether the message is stored. Pub/sub broadcasts a message immediately to all subscribers while message queue batches each message until being processed and deleted by a consumer. Pub/sub provides many-to-many communications while message queue is one to one.

Implementation mode can be distributed or centralized. *Distributed* implementation needs no central node to control operation and communication of the other nodes. All nodes are on an equal ground for peer-to-peer communications. Some protocols require a central point such as a server or a broker, which accepts all messages and then delivers them to intended recipients.

Transport layer provides end-to-end communication with performance features in general. As described in Section 1.1, the IP stack offers two transport protocols: TCP and UDP. TCP guarantees reliability while UDP performs well in speed.

Payload specifies the format and capacity of the actual information carried in a message. Therefore, a protocol's payload dictates how the protocol integrates with a system. For example, web services and technologies benefit from an HTTP protocol that supports various payloads: HTML (HyperText Markup Language), plain text such as ASCII (American Standard Code for Information Interchange) characters, coded text such as UTF-8 (UCS Transformation Format 8), and XML (Extensible Markup Language).

Service discovery finds out who and what on the network before communicating with them. A protocol with service discovery capability offers means by which devices can recognize peers or associated group members in the network. Content distribution networks, such as Netflix and YouTube, deploy service discovery protocol to select specific servers for streaming stored videos.

Device management configures and controls devices.

QoS defines the level of performance that a protocol delivers. Metrics include reliability (such as no-loss), delay, and throughput.

Security and Privacy ensure a system protects data stored/exchanged with Confidentiality from breaching, Integrity from fraudulence, and Availability immunized against DoS attacks (the C.I.A. security goals) as well as a user's right to control the access of one's personal information (privacy).

The rest of the section describes some dominant Communication Protocols with the basic characteristics defined above, but note that not all features are taken by each protocol. We focus on the use cases of M2M communications.

4.1. *Request/response interaction*

HTTP forms the backbone of the World Wide Web. IETF specifies HTTP, including its message formats, and evolves versions: RFC 1945 HTTP/1.0 (1996), RFC 7230 HTTP/1.1 (2014), and RFC 7540 HTTP/2 (2015). HTTP is implemented in two programs: a client by a Web browser (e.g., Google Chrome and Microsoft Edge) and a server to house Web objects such as HTML files or JPEG images (e.g., Apache and Microsoft Internet Information Services). HTTP defines how clients request Web pages from servers and how servers respond to clients, using TCP as the underlying transport protocol. Therefore, HTTP has two types of messages: *HTTP Request Message* with a choice of different methods (e.g., GET and POST) and *HTTP Response Message* with three sections (Status line, Header lines, and Entity body). Refer back to Section 1.2 about a GET message illustrated in Figure 3.

HTTP extends many value-added services. As specified in RFC 6265, HTTP uses cookies to compensate HTTP server's statelessness for sites to keep track of users. In RFC 7234, HTTP deploys proxy servers for Web caching to balance loads and speed up user-perceived responses.

REpresentational State Transfer (REST), known as RESTful API, is an architectural style for distributed hypermedia systems. REST is not HTTP but a paradigm to design web services so that systems communicate with each other easily. RESTful systems are characterized as stateless and separation of concerns between client and server.

Constraint Application Protocol (CoAP) is another standardized request/response protocol. As stated in RFC 7252, CoAP redesigns functions of HTTP for M2M applications, such as energy automation, with resource-constrained nodes and networks. Constrained nodes, like embedded devices with 8-bit microcontrollers, have limited resources of processing, storage, and bandwidth. Constrained networks, like IPv6 over Low-Power Wireless Personal Area Networks (6LoWPANs) [RFC 4944] with a typical data rate of 10 kbps, are low-power, small-throughput, and high-lossy networks. CoAP provides a request/response interaction between application endpoints, similar to HTTP client/server model, but M2M interactions make CoAP devices act in both client and server roles. It supports discovery of services and resources. CoAP easily integrates with HTTP by keeping the key Web concepts including URL, media types, and methods like GET and POST while meeting the requirements of constrained environment with low overhead and simplicity. Unlike HTTP, CoAP uses UDP as its transport protocol with optional reliability supporting unicast and multicast requests. As shown in Figure 13, CoAP uses two logical layers in a single protocol. Its layer of messages complements UDP with reliability. Four types of messages exist: Confirmable (CON), Non-confirmable (NON), Acknowledgment (ACK), and Reset exchange orthogonal to the request/response interactions. Its layer of Requests/Responses interacts with application endpoints using Method and Response Codes. The two logical layers are implemented as just features of the CoAP header.

A message not requiring reliable transmission can be sent as a *NON* message. Such a message, although not acknowledged, still carries a Message ID for duplicate detection. When a recipient cannot process a NON message, it replies with a *Reset (RST)* message. Unreliable use cases include single measurements from a stream of sensor data.

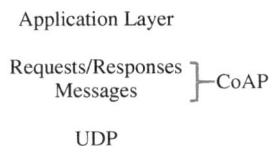

Application Layer

Requests/Responses ⎤
     Messages   ⎦—CoAP

UDP

Figure 13. Abstract layering of CoAP [RFC 7252].

A message requiring reliability is marked with *CON*. It gets retransmitted using timeout and exponential back-off between retransmissions until receiving an *ACK* message with the same Message ID from the corresponding endpoint. When a recipient cannot process a CON message or even provide a suitable error code, it replies with RST instead of ACK.

Binding to Datagram Transport Layer Security (DTLS) [RFC 6347], CoAP defines several security modes, ranging from no security to certificate-base security. A bootstrap process provides a CoAP device with the security information including keying materials and access control lists. The device then sets at one of four security modes: *NoSec* that disables DTLS as if no security (since alternative techniques provide lower-layer protection such as IPsec or link layer security); *PreSharedKey* that enables DTLS and provides a list of pre-shared keys [RFC 4279] with the associated nodes; *RawPublicKey* that enables DTLS, validates an asymmetric key pair without a certificate (a raw public key) using an out-of-band mechanism [RFC 7250], calculates its identity from its public key, and provides a list of the nodes it can communicate with; and *Certificate* that enables DTLS, sets an asymmetric key pair with an X.509 certificate [RFC 5280] that binds it to its subject and is signed by some common trust root, and provides a list of root trust anchors that can be used for validating a certificate.

The IETF Constrained RESTful Environments (CoRE) Working Group standardizes CoAP and its added functions. For example, RFC 8323 supplements CoAP transport protocol choice over TCP, TLS (Transport Layer Security), and WebSockets for enhanced reliability, security, and programmability. RFC 8974 extends CoAP with the extended token length (TKL) field in the CoAP message header for stateless clients and intermediaries to keep per-request state.

4.2. *Publish/subscribe (pub/sub) messaging*

Historically, pub/sub messaging came from the need for data modeling languages associated with network management protocols. *SNMP*, defined by RFC 1157 SNMPv1 (1990), RFC 1901 SNMPv2

(1996), and RFC 3410 SNMPv3 (2002), was the first Internet standard for network management. Its associated data modeling language was *Structure of Management Information (SMI)* RFC 1155 and RFC 2578, which describes *Management Information Base (MIB)* RFC 1066 and RFC 1213. RFC 3780 *SMIng, Next Generation SMI*, defines an experimental protocol as lessons learned from the attempt to decouple a syntactic structure from the management protocol SNMP. *Yet Another Next Generation (YANG)*, specified in RFC 6020, is a data modeling language for *Network Configuration Protocol (NETCONF)* [RFC 6241] (that obsoletes RFC 4741). NETCONF provides mechanisms based on XML encoding to install, manipulate, and delete the configuration of network devices. YANG models data in an XML tree format for NETCONF protocol, including configuration, state data, remote procedure calls (RPCs), and notifications. As specified in RFC 7407, YANG also works with the SNMP agent configured by NETCONF. The inefficiency of periodic fetching or polling YANG datastores suggests the need of pub/sub service for YANG datastore updates. RFC 7923 provides *Requirements for Subscription to YANG Datastores*. *Common Object Request Broker Architecture (CORBA)*, a standard by Object Management Group (OMG), defines the mechanism of Object Request Broker by which objects transparently make requests to and receive responses from each other on the same machine or across a network, showing precursor of pub/sub. Two generalized pub/sub implementations, to be discussed below, demonstrate a new pub/sub technology highly scalable for a distributed datastore connecting millions of edge devices.

Extensible Messaging and Presence Protocol (XMPP), originally named *Jabber* for a text-based messaging protocol, is an application profile of XML enabling near-real-time exchange of structured yet extensible data between devices over a network. RFC 6120 defines the core protocol methods of XMPP: setup and teardown of XML streams, channel encryption, authentication, error handling, and communication primitives for messaging, network availability, and request–response interactions. XMPP uses TCP as its transport protocol, yielding reliable but slow responses. XMPP architecture resembles *Simple Mail Transfer Protocol (SMTP)* for emails [RFC 2821] (that obsoletes RFC 821, 974, 1869): Clients do not talk directly to one another as it is decentralized, and anyone can run a server while there is no central authoritative server. This federated open system allows clients to interoperate with each other on any server using global addresses called "JID" user accounts, similar to email addresses with a username and a domain name separated by an at sign – @. For example, JID <Alice@im.com> is user Alice registered at server <im.com>. A user may log in from multiple locations by specifying a resource, such as work or mobile, with a slash symbol. For example, JID <Bob@example.net/mobile> is Bob's mobile account at <example.net>. XMPP involves concurrent information transactions in a client-to-server or server-to-server session. As shown in Figure 14, the client <Alice@im.com> associated with the server <im.com> exchanges data with the client <Bob@example.net/mobile> associated with the server <example.net>. Therefore, end-to-end communication in XMPP is logically peer-to-peer but physically client-to-server-to-server-to-client.

Two fundamental concepts of XMPP enable any networked entities to make a rapid asynchronous exchange of small structured data chunks: XML streams and XML stanzas. *XML stream* is a container for the exchange of XML elements, which starts with an opening XML <stream> tag and ends with a closing XML <stream> tag. The initiating entity (a client or server) negotiates with the receiving entity (a server) with an "initial stream" viewed as a unidirectional connection. To send data in the opposite direction, the receiving entity must negotiate a "response stream" with the initiating entity. During a stream's lifetime, an entity can send an unbounded number of XML elements over the stream. XML elements include stream negotiation (e.g., TLS negotiation) or XML stanzas. *XML stanza* is

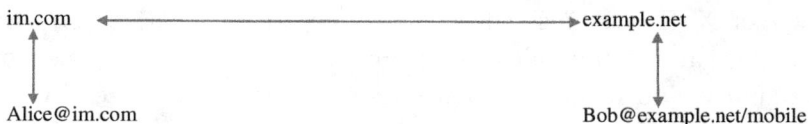

Figure 14. Distributed client-server architecture.

a first-level element, at depth $= 1$ of a stream, as the basic semantic unit. An XML stanza typically contains child elements (with accompanying attributes, elements, and XML character data) to convey the desired information. There are three kinds of stanzas: "message", "presence", and Info/Query or "IQ" for short. They reflect three communication primitives: "push" mechanism for generalized messaging or "message", "pub/sub" mechanism for broadcasting information about network availability or "presence", and "request/response" or "IQ" mechanism for more structured information like HTTP.

XMPP is designed for instant messaging, presence information, and contact list maintenance. Its features such as naming federation, pub/sub, authentication, and security even for mobile endpoints have been explored for IoT. The XMPP Standards Foundation develops extensions to XMPP in its XEP series, including *XEP-0381 Internet of Things Special Interest Group (IoT SIG)*.

Data Distribution Service (DDS) is an M2M pub/sub protocol for real-time systems, standardized by OMG. Nodes that produce information, called *publishers*, create a "topic" (e.g., temperature) and publish "samples". Nodes that declare an interest in that topic, called *subscribers*, receive the samples delivered by DDS. Any node can be a publisher, subscriber, or both simultaneously. Acting as a middleware, DDS provides a standard API for data-centric connectivity. *Middleware* is the software lying between applications and the operation system in a distributed system. As shown in Figure 15, the **Middleware** hides the **Platform** details of the operating system, network transport, and lower-layer data formats while providing the particular **Application** with API in different programming languages to exchange information requiring QoS, security, discovery, and more managed by the Middleware. *API* along with its libraries simplifies the development of distributed systems by making software developers focus on the specifics of their application rather than the mechanics of information exchange. *Data centricity* is unique to DDS. Traditional message-centric middleware, such as CORBA discussed in the beginning of this subsection, expects programmers to manage the complexity of message exchange in their application code.

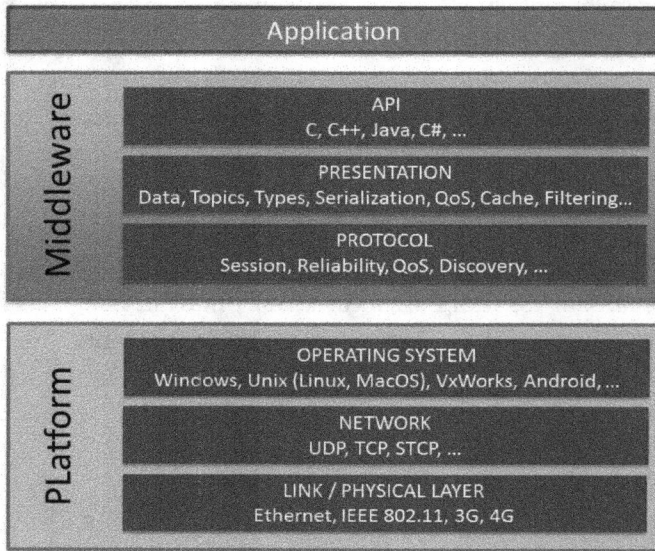

Figure 15. DDS middleware [OMG].

Programmers using a data-centric middleware simply write code to specify when to share data, letting DDS deal with what data to store and how to control the shared data.

DDS uses its own interoperability protocol called *Wire*, which enables different implementations of DDS to communicate smoothly. It offers an extensive QoS support for efficient and versatile realizations of QoS policies, including reliability, bandwidth, delivery deadlines, and resource limits. Such flexible QoS stems at DDS allowing TCP, UDP, Shared-Memory or other transport protocols. DDS also supports fine-grained security (Ashok & Edgar, 2021).

DDS adopts Interface Definition Language (IDL) [OMG IDL4] [ISO/IEC 19516:2020] to define data types and interfaces, independent of the programming language. Furthermore, a Unified Modeling Language (UML) profile specifies DDS domains, topics, and how to publish and subscribe objects without first describing the types in another language such as XML. Notice that DDS and CORBA are two independent connectivity standards by OMG, each supporting different use cases and integration patterns: DDS for pub/sub data distribution (including RPC) while CORBA for remote method invocation on distributed objects. Since both leverage the same IDL, applications can use the same data types for both DDS and CORBA interactions.

Highly scalable architecture enables DDS systems spanning from Edge to Fog to Cloud: Machine Domain (Edge), Control and Central Domains (Fog), and Admin Domain (Cloud). Low-latency data connectivity, reliability, security, interoperability, and scalability across thousands even millions of devices make DDS suitable for mission-critical IoT applications (Dahlmanns *et al.*, 2021).

4.3. *Message queue operation and device management*

Message Queuing Telemetry Transport (MQTT) protocol was first drafted by IBM and Arcom (now Cirrus Link) to collect data from remote devices backing up the server for minimal battery loss and constrained bandwidth. The Organization for the Advancement of Structured Information Standards (OASIS) and International Organization for Standardization (ISO)/International Electrotechnical Commission (IEC) standardized MQTT [OASIS v3.1.1: 2014] [ISO/ IEC 20922: 2016], changing its focus from proprietary embedded systems to open IoT use cases. MQTT is no longer a message queuing protocol; MQTT is a client–server pub/sub messaging transport protocol for lightweight and open-source M2M communications. To clarify the confusion, we discuss MQTT in this subsection for Device Management, instead of the previous subsection about Pub/Sub Messaging.

The MQTT pub/sub message pattern provides one-to-many message distribution and decoupling of applications. Small transport overhead and minimized protocol exchanges reduce network traffic. The size of a Fixed Header is just 2 Bytes. MQTT is suitable for resource-constrained devices and bandwidth-limited networks, like IoT contexts where a small code footprint is required. MQTT relies on TCP as the transport protocol for reliable connections.

Note: A variant, MQTT for Sensor Networks (MQTT-SN), uses non-TCP/IP networks such as UDP or ZigBee for WSN. MQTT defines three QoS levels in increasing order of protocol processing overhead:

- "at most once" (0) delivers messages by the best efforts, so loss might occur.
- "at least once" (1) assures messages' arrival, so duplicate might occur.
- "exactly once" (2) assures each message to arrive exactly once.

There are two types of network entities in MQTT — **MQTT Broker** (a server) and **MQTT Clients** — connected to the broker to send and receive messages on the channels of their interest, as shown in Figure 16. Sending clients are *Publishers* and receiving clients are *Subscribers*. A publisher does not need any information about subscribers such as their locations, and subscribers, in turn, do not have to know publishers. It is the broker that distributes the messages sent by publishers to any subscribers on specific topics. If a broker receives a message on a topic with no subscribers, the broker discards the message unless the publisher designated the message as a retained message so that clients receive the retained message once they subscribe.

https://mqtt.org/assets/img/mqtt-publish-subscribe.png

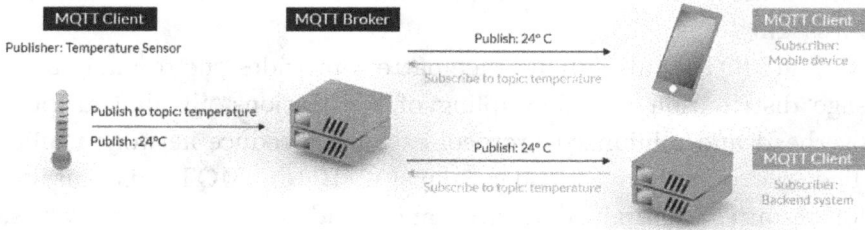

Figure 16. MQTT pub/sub architecture [mqtt.org].

MQTT is a bidirectional communication protocol, so a client can both send and receive messages by both publishing and subscribing. It facilitates functions of sharing data and managing devices. Multiple brokers can be configured for automatic backup in case of broker failure to balance load.

Regarding security, MQTT uses TLS encryption with username and password. OASIS provides a supplemental publication: MQTT and the NIST Framework for Improving Critical Infrastructure Cybersecurity [MQTT NIST]. IETF has been recently drafting an MQTT-TLS profile of Authentication and Authorization for Constrained Environments (ACE) Framework 2022.

Advanced Message Queuing Protocol (AMQP), rooted in the financial industry, has been standardized by [OASIS v1.0: 2012] and [ISO/IEC 19464:2014]. It relies on queues between senders and receivers for reliable delivery of messages. AMQP uses TCP as the transport protocol to implement the reliability. Unlike HTTP where a connection has to be initiated by a client, AMQP allows a server to initiate a connection with a client. AMQP is flexible, offering peer-to-peer, client-to-broker, and broker-to-broker communications. AMQP.org, a forum by financial services and technology providers, claims AMQP as the IP for business messaging. As shown in Figure 17, AMQP infrastructure enables an open, secure, reliable, and standardized ecosystem for commoditized multi-vendor communications to conduct business on the Internet and in the Cloud.

AMQP layered architecture contains three major components: Network Protocol, Message Representation, and Broker Services.

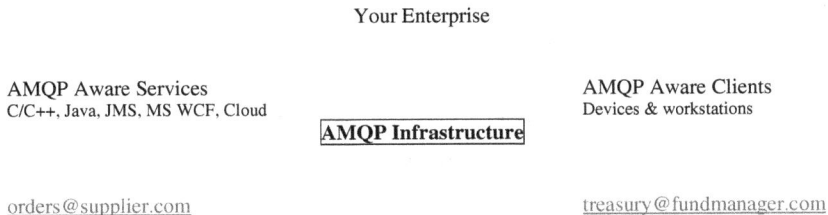

Your Enterprise

AMQP Aware Services
C/C++, Java, JMS, MS WCF, Cloud

AMQP Infrastructure

AMQP Aware Clients
Devices & workstations

orders@supplier.com

treasury@fundmanager.com

Figure 17. Business partners and services [amqp.org].

AMQP Network Protocol is a peer-to-peer protocol featuring reliability, capability, and security. Normally, one peer acts in a client application role while the other peer plays a trusted broker role of message queuing, routing, and delivery. The protocol multiplexes a TCP connection for multiple conversations, simplifying firewall management in addition to controlling flows from producers by consumers and maintaining the life cycle of a message through fetching, processing, and acknowledgment. It contains mechanisms to resume message transfers in the event of lost connections. Its messaging capabilities include discovering sources of messages for a specific topic, pre-fetching messages to enhance performance, and processing batches of messages within a transaction. It provides comprehensive security mechanisms, such as TLS and Kerberos for seamless end-to-end confidentiality.

AMQP Message Representation envelopes data for portability among systems accessible by many programming languages: C/C++, Python, Java, Java Message Service (JMS), and Microsoft Windows Communication Foundation (WCF) to name a few. Its Type System and Message Coding facilitate the portability like the "envelope" of the message to add routing properties while hiding the low-level network protocol from users of messaging software.

AMQP Broker Services define the basic semantics of using a message broker to send messages transparently without dealing with technical details. The basic unit of data in AMQP is *frame*. An open frame by a sending peer initiates a *connection* between two peers, which is terminated with a close frame. Over a connection, several sessions can be multiplexed. Each *session*, initiated with a begin frame and terminated with an end frame, facilitates a bidirectional sequential conversation between two peers. In a session, multiple links, in both directions, can be grouped together. A *link*, initiated with an attach frame and terminated with a detach frame, is established to send or receive messages. Messages are sent on a link unidirectionally by using a transfer frame, controlled with flow frames for a receiver to protect itself from being overwhelmed by a large volume of messages or to pull messages on a subscribed link, and

```
                                        Bare Message
                                             |
                        .--------------------+--------------------.
                        |                                         |
                        |                                         |
+--------+-----------+-----------+----------+-----------+-----------+--------+
| header | delivery- | message-  | properties | application- | application- | footer |
|        | annotations | annotations |          | properties | data       |        |
+--------+-----------+-----------+----------+-----------+-----------+--------+
|                                                                          |
'--------------------------------------------------------------------------'
                                        |
                                Annotated Message
```

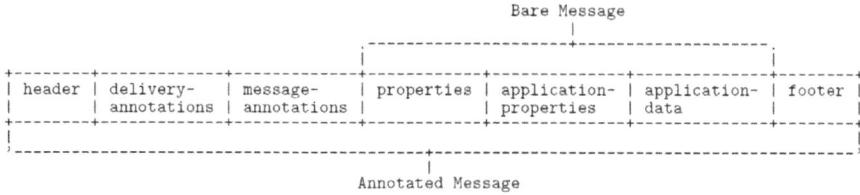

Figure 18. AMQP message format [OASIS v1.0: 2012].

settled between the sender and receiver with disposition frames to guarantee reliability. Like MQTT, message delivery in AMQP guarantees three QoS levels: "at most once", "at least once", and "exactly once".

AMQP clarifies the confusing term message viewed by senders and receivers with "bare message" as the immutable payload supplied by the sender and "annotated message" as the bare message plus annotation sections along with the header and footer supplied by the messaging infrastructure along the way finally seen by the receiver. Figure 18 summarizes the AMQP message format. Annotations are of two classes: *delivery annotations* that are consumed by the next node along the route and *message annotations* that travel with the message indefinitely. Bare message has three sections: *properties* of standard, *application properties*, and *application data* called "body". Each of the six non-body sections can be zero or one. The body chooses one of the three cases: one or more data sections, one or more AMQP-sequence sections, or a single AMQP-value section.

The *header* carries standard delivery details about the transfer of a message through the AMQP network. If omitted, the receiver assumes the default values. The *delivery annotations* convey the information about delivery-specific non-standard properties from the sending peer to the receiving peer, negotiated on link attach. If omitted, the receiver takes an empty map of annotations. The *message annotations* convey the information about the message properties at the infrastructure; thus, intermediaries must propagate the annotations across every delivery step. The capabilities are also negotiated on link attach. If omitted, the receiver also takes an empty map of annotations. The *footer* carries details about the message or delivery which can only be calculated or evaluated once the whole bare message has been constructed. Example footers are message hashes, signatures, and encryption details.

As mentioned before, the bare message is immutable; thus, if an intermediary retransmits a bare message, it must not alter the message. The *properties* section defines a set of standard attributes of the message, including the identity of the user responsible for producing the message (set by the client and possibly authenticated by intermediaries), the address of the destination node, and a subject to summarize the message content and purpose. The *application properties* section provides structured application data for intermediaries to apply filtering or routing functions. The *application data* section contains opaque binary data (data sections), or arbitrary numbers of structured data elements (AMQP-sequence sections), or a single AMQP value.

AMQP provides a set of standardized but extensible messaging capabilities. For example, group interactions within atomic transactions by *Transactional Messaging*. The coordination extends to any number of transfers spread across many distinct links in either direction.

5.   WAVE for Connected Vehicles: A CPS Example

As a kind of CPS (Cheng, 2014), Connected Vehicles (CVs) or the Internet of Vehicles (IoV) enables safe, efficient, and convenient road transportation using interoperable networked wireless communications among vehicles, infrastructure, and personal devices. According to USDOT, some automotive manufacturers have deployed connected vehicle technology for light vehicles with no more than $350 per vehicle in 2020, Cadillac models by General Motors, for example.

Related to but different from CVs, Automated Vehicles (AVs) use a host of on-board sensors, cameras, and radar applications for the same purpose, at six levels of driving automation defined by the Society of Automotive Engineers (SAE) International [SAE J3016].

SAE J3016™ LEVELS OF DRIVING AUTOMATION™
Learn more here: sae.org/standards/content/j3016_202104

	SAE LEVEL 0™	SAE LEVEL 1™	SAE LEVEL 2™	SAE LEVEL 3™	SAE LEVEL 4™	SAE LEVEL 5™
What does the human in the driver's seat have to do?	You **are** driving whenever these driver support features are engaged – even if your feet are off the pedals and you are not steering			You **are not** driving when these automated driving features are engaged – even if you are seated in "the driver's seat"		
	You must constantly supervise these support features; you must steer, brake or accelerate as needed to maintain safety			When the feature requests, you must drive	These automated driving features will not require you to take over driving	
	These are driver support features			*These are automated driving features*		
What do these features do?	These features are limited to providing warnings and momentary assistance	These features provide steering OR brake/ acceleration support to the driver	These features provide steering AND brake/ acceleration support to the driver	These features can drive the vehicle under limited conditions and will not operate unless all required conditions are met		This feature can drive the vehicle under all conditions
Example Features	• automatic emergency braking • blind spot warning • lane departure warning	• lane centering OR • adaptive cruise control	• lane centering AND • adaptive cruise control at the same time	• traffic jam chauffeur	• local driverless taxi • pedals/ steering wheel may or may not be installed	• same as level 4, but feature can drive everywhere in all conditions

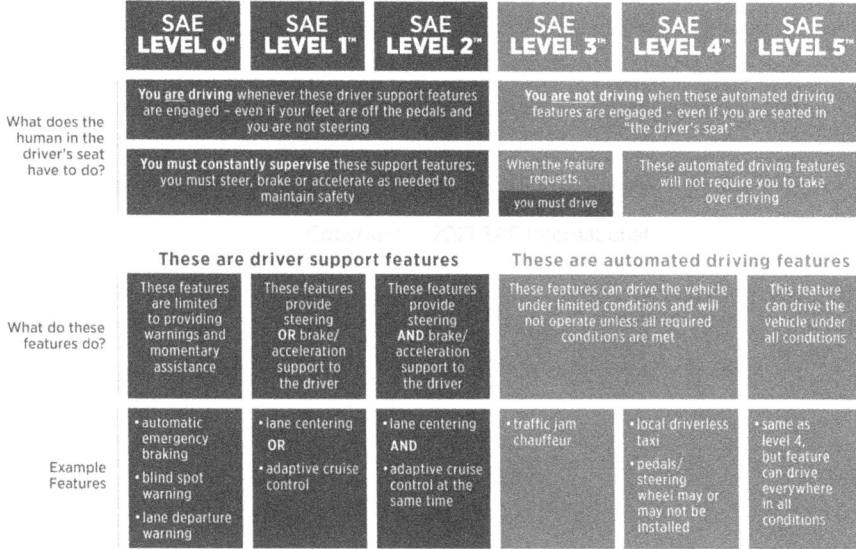

Figure 19. SAE's six levels of driving automation [SAE J3016].

As shown in Figure 19, the lower three levels use driver support systems that need a human driver's constant supervision; the higher three levels let the automated system monitor driving from Level 3 Conditional Automation (a human driver takes over when the system requests) to Level 5 Full Automation (the system can drive the car under all situations).

CV provides additional resources to handle impending dangers with vehicle-to-vehicle (V2V) communications and extend transportation efficiency with vehicle-to-infrastructure (V2I) communications. By transmitting messages about cars' speed, heading, brake status, and other information, CV lengthens detection range and perceives threats beyond line of sight, capabilities that cannot be addressed by current sensor technology. Driverless cars by integrating Connected and Automated Vehicles would drastically improve road safety, expand transportation capability, and extend mobility to everyone like the disabled, elderly, and youth.

This section takes CV as a CPS example to examine how network and communication protocols cooperate for complicated information exchanges in wireless and mobile environments. The lack of ubiquitous high-speed wireless mobile communications and the lack of homogeneous interfaces among automotive manufacturers have limited the provision of externally driven services to vehicles. DSRC addresses the former issue. Section 3 already discussed DSRC and IEEE 802.11p now incorporated into the IEEE 802.11 base standard. The IEEE 1609 Family of Standards for WAVE addresses the latter issue and provides a foundation for the organization of management functions and the modes of system operations. Together, these standards support a broad range of applications for road transportation including vehicle safety, enhanced navigation, automated tolling, and traffic management.

The IEEE 1609 family of standards meets the communication needs of mobile and fixed elements in the transportation environment aligned with CV. The following ITS standards, particularly by SAE International and ASTM (American Society for Testing and Materials) International, should be considered:

- SAE J2354 Message Sets for Advanced Traveler Information System (ATIS)
- SAE J2735 V2X Communications Message Set Dictionary
- SAE J2945 DSRC Systems Engineering Process Guidance/ Common Design Concepts
- ASTM E2213-03 Standard Specification for Telecommunications and Information Exchange Between Roadside and Vehicle Systems — 5 GHz Band DSRC, MAC, and PHY Specifications.

Additionally, organizations involved in ITS aspect/communications or as stakeholders include the following:

- *ISO*
- *Internet Engineering Task Force (IETF)*
- *USDOT National Highway Transportation Safety Administration (NHTSA)*

- *Crash Avoidance Metrics Partners (CAMP)*
- *OmniAir Consortium*, an industry association promoting interoperability and certification for CVs.

5.1. WAVE protocol stack

The IEEE 1609 family of standards for WAVE defines an architecture and a complementary set of services and interfaces that collectively enable secure V2V and V2I communications. As shown in Figure 20, the WAVE family consists of six standards and relies on IEEE 802.11 for MAC and PHY layers:

- *IEEE 1609.0 WAVE Architecture* describes the WAVE architecture and services for multi-channel DSRC/WAVE devices to communicate in a mobile vehicular environment.
- *IEEE 1609.2 WAVE Security Services for Applications and Management Messages* specifies secure message formatting and exchange processing. It may provide additional security services to higher layers.

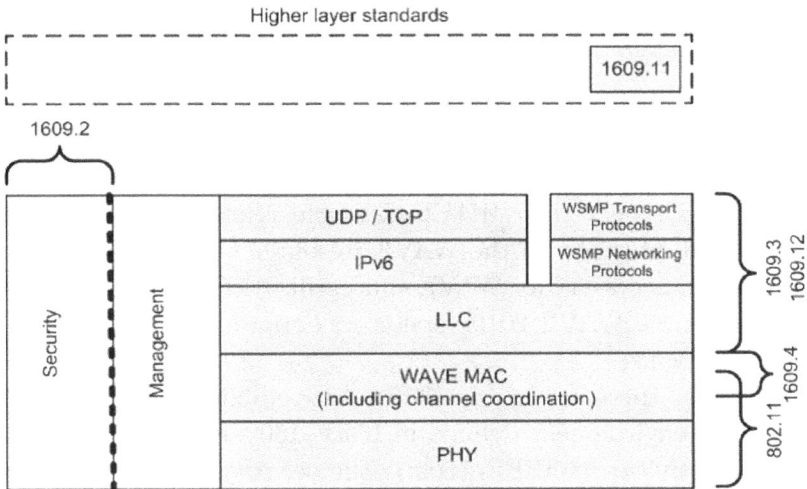

Figure 20. WAVE protocol stack [IEEE 1609.0].

- *IEEE 1609.3 WAVE Networking Services* specifies network and transport layer services. Its features include addressing and routing, WAVE data exchange, WAVE Service Advertisements (WSA) transmission/monitoring and channel access assignment, WAVE Short Message Protocol (WSMP), use of local link control (LLC) sub-layer and Ether-Type Protocol Discrimination (EPD), and streamlined IPv6 configuration. Furthermore, this standard defines the MIB for the WAVE protocol stack.
- *IEEE 1609.4 WAVE Multi-Channel Operations* specifies extensions to IEEE 802.11 MAC layer protocol to support WAVE operations. Its features include channel coordination and routing, multi-channel synchronization, use of IEEE 802.11 facilities, and MAC layer re-addressed in support of pseudonymity (a mechanism for privacy). It also maintains MIB.
- *IEEE 1609.11 Over-the-Air Electronic Payment Data Exchange Protocol for ITS* specifies the services and secure message formats to support secure electronic payments. It is the first application layer IEEE 1609 standard. Its use cases include electronic fee collection.
- *IEEE 1609.12 Identifiers* records the allocations of identifiers used by the WAVE standards. The identifiers include object identifier (OID), Ether Type, and Management ID. It also defines the usage and encoding rules of Provider Service Identifier (PSID).

Note: *IEEE 1609.1-2006 WAVE Resource Manager*, specifying the services and interfaces of the WAVE Resource Manager application, was found unnecessary for WAVE standardization and has been withdrawn since the WAVE 2019 version was approved by the IEEE-SA Standards Board.

Together, these standards define how applications function in the WAVE environment defined in IEEE 1609.0 based on the management activities in IEEE 1609.1, the security protocols in IEEE P1609.2, and the network layer protocol in IEEE 1609.3. They also provide extensions of the physical channel access defined in IEEE 802.11 to support the WAVE standards in IEEE 1609.4.

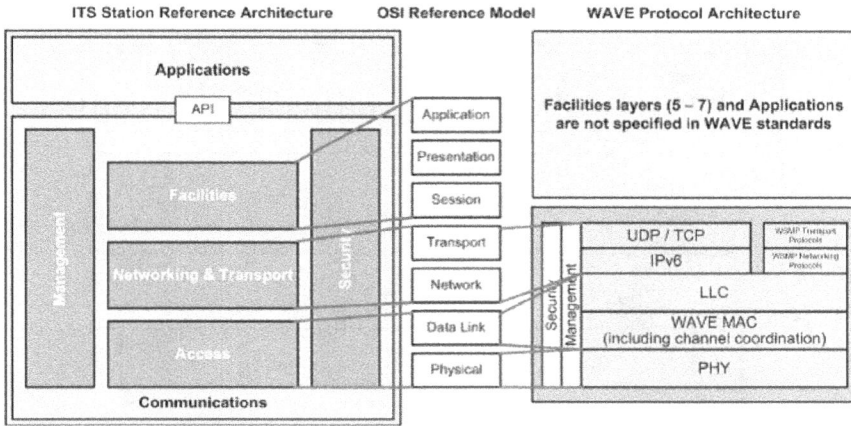

Figure 21. Relationship among protocol models [IEEE 1609.0].

Figure 21 maps the relationship among three standardized models: ITS Station Reference Architecture, ISO *Open Systems Interconnection (OSI) Reference Model*, and IEEE WAVE Protocol Architecture. Derived from the OSI seven-layered model for communications [ISO/IEC 7498-1: 1994], [ISO 21217: 2020] specifies *ITS Station and Communication Reference Architecture*, where Road Safety is one of **Applications** above the OSI's Application layer. **Facilities**, for example, Local Dynamic Map (LDM) to keep track of nearby objects including vehicles, match to the top three layers of the OSI Reference Model: Application, Presentation, and Session. However, the IEEE WAVE family does not specify the OSI's top three layers.

Differences among ITS, ISO, and IEEE have been addressed by international harmonization efforts to align the standards. IEEE WAVE protocols have incorporated their suggested actions in the latest revisions.

5.2. *WAVE system composition*

Figure 22 depicts a device, with two radio transceivers (two sets of MAC/PHY layers), operating on IEEE WAVE Protocol Architecture with two options (IPv6 or WSMP) that supports four applications:

Higher layer (four examples shown)

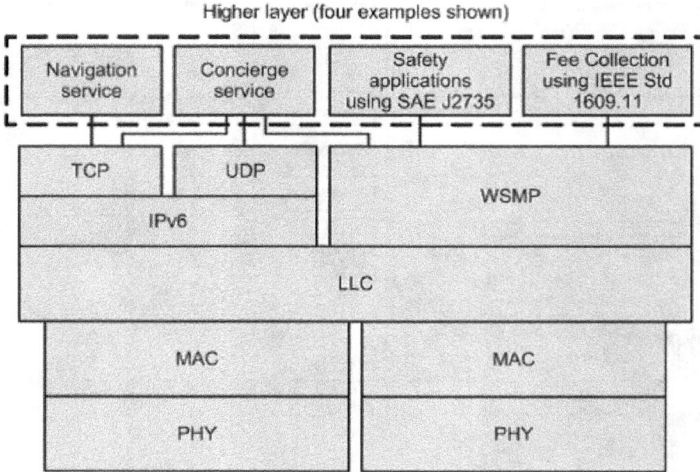

Figure 22. Example WAVE device [IEEE 1609.0].

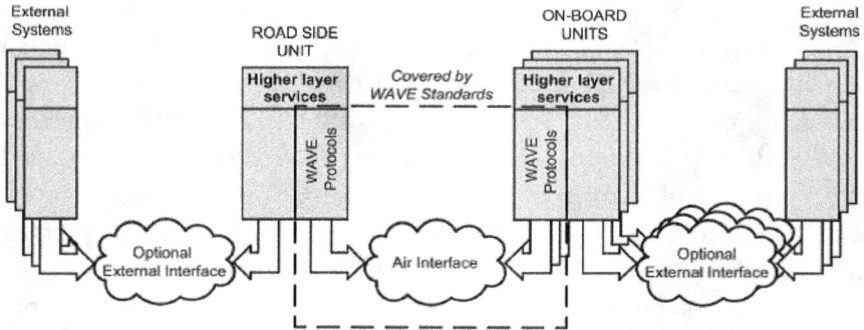

Figure 23. WAVE system composition [IEEE 1609.0].

Navigation service, Concierge service, Safety applications using SAE
J2735, and Fee Collection using IEEE 1609.11.

The primary architectural components of a WAVE system are
On-Board Unit (OBU) installed in a vehicle, **Road-Side Unit
(RSU)** permanently mounted, and **WAVE interface**. As shown
in Figure 23, a WAVE system is composed of a stationary RSU
connected to the Internet, many mobile OBUs communicating among
themselves and with the RSU via Air Interface, and optional External Systems such as pedestrian units.

Applications exchange data in a consistent, interoperable, and timely manner facilitated by WAVE protocols. WAVE devices play one of two roles: **Provider** that advertises its services and **User** that chooses to participate in the advertised service opportunities. Though the roles are not tied to specific device types, an RSU plays Provider in most cases. Similar to Software Defined Network (SDN), two separate planes facilitate WAVE communications: Data Plane and Management Plane.

A wide range of stakeholders are involved in this family of standards besides transportation agencies, automotive and traffic engineers who design, specify, implement, and test WAVE devices. Network engineers, hardware makers, and application designers supporting ITS use these standards to define the communications architecture for V2V and V2I interactions, and to design low-latency interfaces of OBUs and RSUs. ITS application developers follow the standards in implementing seamless applications without regard to specific vendors.

5.3. *Collision avoidance: A WAVE use case*

Collision avoidance by WAVE devices exchanging messages is one of the most important use cases for a WAVE system. Example applications include Forward Collision Warning, Blind Spot Warning, Intersection Violation Warning, Icy Road Warning, and Longitudinal Collision Risk Warning. Figure 24 shows the standards used for collision avoidance applications.

SAE J2735 specifies *Basic Safety Message (BSM)*, *Signal Phase and Timing (SPAT)*, and *Map Data (MAP)* messages. OBUs periodically send their basic state information using BSM, which conveys the sender's position, speed, acceleration, heading, and more. RSUs send additional information using SPAT and MAP messages, which convey the signal state and geographical description of an intersection.

A sender, OBU or RSU, uses IEEE 1609.2 to sign and authenticate the messages transmitted for collision avoidance applications. Unlike the mandatory message integrity by digital signature, encryption is unnecessary for these messages as they are often broadcast.

Example application standards for Safety Applications

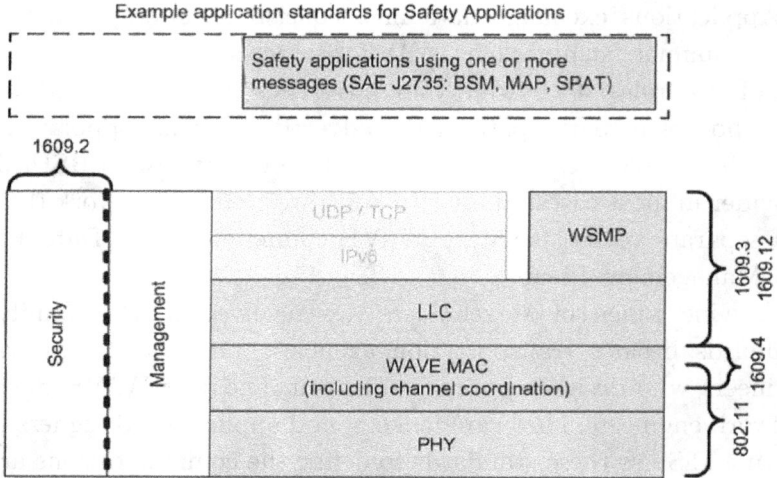

Figure 24.   Protocols for collision avoidance [IEEE 1609.0].

WSMP specified in IEEE 1609.3 transports the signed messages. The choice of WSMP, instead of IPv6, enables control power on a message-by-message basis. WSMP also offers bandwidth efficiency to prevent congestion by a large quantity of safety broadcasts (Ahmadi, Abdelzaher & Gupta, 2010). BSMs and SPAT messages are sent on predetermined channels, so WSA is unnecessary. Their contents are updated as frequently as 10 times per second per sender. MAP messages can choose advertised or predetermined, but the U.S. model is considered to transport all three types of messages (BSM, SPAT, and MAP) on a designated "safety channel" that WAVE devices continuously monitor. Therefore, none of the safety messages need WSA. Similarly, the channel-switching function of IEEE 1609.4 is unnecessary. IEEE 1609.12 allocates three PSID values for applications related to BSM. Two are for vehicle BSMs, one of which is intended for messages composed with vehicle data of high accuracy. The third PSID value is for rail vehicles such as trains. Each PSID value is carried in a WSMP header so that a receiver can efficiently distinguish the three classes of BSM. IEEE 1609.12 also allocates a fourth PSID value for SAE J2735 intersection messages. This fourth PSID value is carried in a WSMP header of an SPAT or MAP message. More PSID values are allocated for other safety applications associated with SAE J2735 messages.

References

Ahmadi, H., Abdelzaher, T.F., and Gupta, I. Congestion control for spatio-temporal data in cyber-physical systems. In *ICCPS '10: Proceedings of the 1st ACM/IEEE International Conference on Cyber-Physical Systems*, pp. 89–98, 2010.

Alexander, A., *et al. Loon Library: Lessons from Building Loon's Stratospheric Communications Service.* Loon LLC, 2021, https://storage.googleapis.com/x-prod.appspot.com/files/The%20Loon%20Library.pdf.

Ashok, A. and Edgar, T. A high-fidelity cyber-physical testbed-based benchmarking dataset for testing operational technology specific intrusion detection systems. In *2021 IEEE International Symposium on Technologies for Homeland Security (HST)*, Boston, MA, USA, 08–09 November 2021.

Barritt, B. and Cerf, V. Loon SDN: Applicability to NASA's next-generation space communications architecture. In *2018 IEEE Aerospace Conference*, pp. 1–9, Big Sky, MT, USA, 03–10 March 2018.

Bhatia, L., *et al.* Control communication co-design for wide area cyber-physical systems. *ACM Transactions on Cyber-Physical Systems*, 5(2):1–27, 2021.

Cheng, A.M. An undergraduate cyber-physical systems course. In *CyPhy '14: Proceedings of the 4th ACM SIGBED International Workshop on Design, Modeling, and Evaluation of Cyber-Physical Systems*, pp. 31–34, April 2014.

Dahlmanns, M., *et al.* Transparent end-to-end security for publish/subscribe communication in cyber-physical systems. In *Proceedings of the 2021 ACM Workshop on Secure and Trustworthy Cyber-Physical Systems (SAT-CPS'21)*, pp. 78–87, Virtual Event, USA, 28 April 2021.

Hill, J.L. and Culler, D.E. MICA: A wireless platform for deeply embedded networks. *IEEE Micro*, 22(6):12–24, 2002.

Jawhar, I., Al-Jaroodi, J., Noura, H., and Mohamed, N. Networking and communication in cyber physical systems. In *2017 IEEE 37th International Conference on Distributed Computing Systems Workshops (ICDCSW)*, pp. 75–82, Atlanta, GA, USA, 05–08 June 2017.

Kovatsch, M., *et al. Web of Things (WoT) Architecture*, 2020 [Online]. Available at: https://www.w3.org/TR/wot-architecture/.

Kurose, J.F. and Ross, K.W. "1.5 protocol layers and their service models". In *Computer Networking: A Top-Down Approach*, 8th ed., pp. 49–52. Pearson: Hoboken, NJ, 2021.

Levis, P., *et al.* TinyOS: An operating system for sensor networks. In *Ambient Intelligence*, pp. 115–148. Springer, Berlin, Heidelberg, 2005.

Liu, C. (H.) and Zhang, Y. *Cyber Physical Systems: Architectures, Protocols and Applications*. CRC Press, Boca Raton, FL 2019.

Mainetti, L., Mighali, V., and Patrono, L., A software architecture enabling the Web of Things. *IEEE Internet of Things Journal*, 2(6):445–454, 2015.

Ma, Y., *et al.* Holistic cyber-physical management for dependable wireless control systems. *ACM Transactions on Cyber-Physical Systems*, 3(1):1–25, 2019.

Rajkumar (Raj), R., Lee, I., Sha, L., and Stankovic, J. Cyber-physical systems: The next computing revolution. In *Design Automation Conference*, pp. 731–736, Anaheim, CA, USA, 13–18 June 2010.

Sciullo, L., *et al.* A survey on the Web of Things. *IEEE Access*, 10:47570–47596, 2022.

Serrano, P., *et al.* Balloons in the sky: Unveiling the characteristics and trade-offs of the Google Loon Service. *IEEE Transactions on Mobile Computing*, pp. 1–14, 2021.

Sha, L., Gopalakrishnan, S., Liu, X., and Wang, Q. Cyber-physical systems: A new frontier. In *2008 IEEE International Conference on Sensor Networks, Ubiquitous, and Trustworthy Computing (sutc 2008)*, pp. 1–9, Taichung, Taiwan, 2008.

Vixie, P., Sneeringer, G., and Schleifer, M. Events of 21-Oct-2002. 24 November 2002 https://web.archive.org/web/20110302164416/http://www.isc.org/f-root-denial-of-service-21-oct-2002 (accessed 20 May 2023).

Wan, J., Lopez, A., and Faruque, M.A.A. Physical layer key generation: Securing wireless communication in automotive cyber-physical systems. *ACM Transactions on Cyber-Physical Systems*, 3(2):1–26, 2018.

Watteyne, T., *et al.* OpenWSN: A standards-based low-power wireless development environment. *Transactions on Emerging Telecommunications Technologies*, 23:480–493, 2012.

Zimmerling, M., *et al.* Adaptive real-time communication for wireless cyber-physical systems. *ACM Transactions on Cyber-Physical Systems*, 1(2):1–29, 2017.

https://doi.org/10.1142/9789811273551_0003

Chapter 3

Forensics in Cyber-Physical Systems

Gökhan Kul* and Chidera Biringa†

Department of Computer and Information Science,
University of Massachusetts Dartmouth,
Dartmouth, MA, USA

*gkul@umassd.edu
†cbiringa@umassd.edu

Abstract: Cyber-Physical Systems (CPSs) can be perceived as a generation of systems that integrate computational and physical capabilities that allow human interaction. This interaction can act as a key enabler of novel technologies such as autonomous driving, critical infrastructure, safety critical systems, and aviation systems through the capabilities that connect the computational and communication aspects of information systems with the physical world. However, these systems are prone to cyberattacks, physical attacks, and also failures of both software and hardware systems. After an accident or a failure, applying forensic techniques on these systems can identify the root cause of the problem. Determining the root cause helps the system owners to fix any vulnerabilities, if the reason of the failure is an intentional attack, or to fix the bugs or problems that eventually led to the failure or accident. In this chapter, we will focus on the components of the CPS and the data curation approach from these components.

Keywords: Digital Forensics, Investigative Process, CPS Analysis.

1. Introduction

Digital forensics is a branch of forensic science concerned with the use of digital information produced, stored, and transmitted by computers as a source of evidence in investigations and legal proceedings.

The research and development of Cyber-Physical Systems (CPS) pose certain challenges that do not manifest commonly in general-purpose computing such as the following: (i) while performance is an important aspect of general-purpose computing, performing a task in a certain time window in CPS can be critical to the operation of the whole infrastructure [1], (ii) the systems commonly use AI/ML pipelines that will need feedback loops to self-correct [2], (iii) to establish situational awareness, the systems require extensive sensor and actuator technologies that capture direct and indirect data from the environment [3], (iv) the commonly accepted hardware/software interface that runs on interrupts may not be as efficient to satisfy the needs of CPS [3], and (v) heterogeneous information and components produce huge volumes of data [4]. There are also other challenges in CPS that manifest themselves in distributed systems such as (i) net-worked connections, (ii) distributed control and synchronization, and (iii) verification and validation issues. All these challenges require sig-nificant effort to tackle and establish to ensure the safety, stability, and performance desired. Both academia and industry seek to mini-mize costs and effort to produce such systems as it would be the case in other domains.

However, due to the complexity, the *attack surface* of CPS is considerably large [5]. Therefore, attackers can exploit vulnerabilities originating from design flaws due to complexity, integration issues, individual component vulnerabilities, compatibility issues, and also human-in-the-Loop.

On the other hand, various components and actors in CPS provide a wide range of opportunities for forensic data and evidence collec-tion. Digital forensics is the science of investigating criminal activity through examination of the environment, scene, history of activities of actors, digital devices and components, and extracting actionable information [6].

In this chapter, we will cover (i) CPS-related work in Section 2, (ii) the role of security and privacy in CPS in Section 3, (iii) investigation processes in Section 4, and (iv) forensic approaches and analysis concerning legal and technical perspectives in Section 5. Lastly, Section 6 concludes this chapter.

2. Related Works

The existing research on the formulation and application of CPS is broad and diverse. Cybersecurity and forensics are usually interdisciplinary domains integrated into other research fields. Cybersecurity gives us the means to protect a system's assets from potentially malicious actors, while forensics details investigative methods used to retrieve or recover compromised information. In both academic literature and industrial systems, we encounter CPS applications mostly in the healthcare and process control fields. CPS utilizes observational data provided by sensors and actuators to enable control and interaction with the physical world. Thus, protecting these assets from malicious actors is central to guaranteeing the confidentiality, integrity, and availability of CPS.

Research in the healthcare field predominantly applies intelligent sensors to monitor patient conditions and to aid in surgical operations. Applications to aid in clinical decision-making via clinical sensors [7], to classify stroke vulnerability among healthy patients via Electrocardiogram wearable devices [8], or to prevent human errors during surgical operations via a robotic scrub nurse [9] are just a few examples of how safety and security are critical to healthcare-related CPS.

Similarly, CPS in process control [10] is widely applied in the production of various products and is the most direct application of CPS, especially in industrial 4.0 manufacturing. Novel process control systems address challenges with time inference and memory utilization in the real world [11], and observe and control how operational sensor data are collected, mined, and uploaded to a central server [12].

There has been a significant degree of research in digital forensics in areas such as malware analysis [13,14], memory evaluation [15,16], network [17,18], mobile data [19,20], and database [21,22] forensics.

The application of digital forensic methods and tools in CPS is interconnected with the components they aim to investigate, as attack surfaces and intrusion points emerge from multiple components in multiple phases. Therefore, research mostly focuses on individual components and their domains. However, several works in forensic research specifically target CPS.

Intrusion detection systems (IDS) and intrusion prevention systems (IPS) form an important part of the security mechanisms in CPS [23]. Dynamic logging of system data [24] such as programmable logic controllers is also prevalent for forensic analysis. These controllers are commonly used in industrial control and supervisory control systems. Efficient logging enforces traceability that aids in identifying and tracking an attacker for forensic purposes. Data from IDS and IPS can be crucial in forensic diagnosis as some systems can use these data such as a dynamically distributed forensic diagnosis and analysis framework [25] for a complex cyberattack. The said framework facilitates the comprehension of extensive and multiphase adversarial attacks. The data can also be useful in the analysis of reoccurring security incidents in CPS [26].

The forensic analysis of the interaction between software and memory is another area that was explored in the CPS domain. Al-Sharif *et al.* [27] presented an exploration and analysis of Java software random-access memory. The authors experimented with forensic evidence recovery on several Java Virtual Machine (JVM) platforms. Their experiment showed that pertinent data of specific segments of a program process state are resistant to total wipes even after the explicit invocation of the garbage collector and the JVM process termination. These forensic traces facilitate data retrieval from an attacker's computer during legal proceedings.

3.   Role of Security and Privacy on CPS Forensics

CPSs are mainly used for optimization and revenue maximization in additive manufacturing, automation, robotics, and the materials industry. Furthermore, they create significant attack surfaces

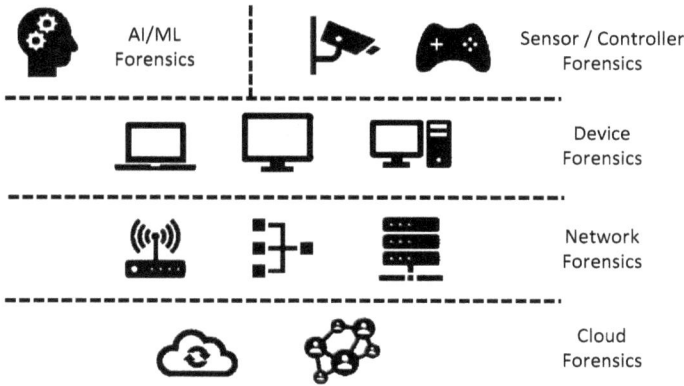

Figure 1. Scope of CPS forensics.

for systematic attacks, industrial or governmental espionage, or cyber-sabotage. In the first half of 2021, hacker groups successfully conducted cyberattacks on critical infrastructure, pipelines, meat manufacturers, and state government systems, resulting in millions of dollars of lost revenue, higher inflation rates, and problems in supply chains. This is a strong indicator that hackers will continue attacking important infrastructure that is considered a CPS (Figure 1).

Cyberattacks have two main categories [28]:

- **Passive attacks:** The attacker observes the system to glean data or the structure of the system, and potentially steal information.
- **Active attacks:** The attacker tampers with the integrity or availability of the system.

In both cases, the forensic investigator will have to investigate each CPS component, which may include nodes deployed in the small scale (such as smart thermostats) to large-scale systems (such as autonomous vehicles).

Since passive attacks do not tamper with the system or the data in any observable way, they tend to only leave traces in logs that concern searching and accessing files and data on the system. They violate the privacy of the stakeholders of the system, and the data they extract may be used later in other attacks, not necessarily in the cyber domain, such as identity theft.

The active attacks may tamper the confidentiality, integrity, or availability of the system, and they can cause wide service disruptions, malfunctions, failure, and destruction of the system. They may also cause physical harm to the environment and to humans considering the CPS domain.

CPSs contain multiple processing units, sensors and actuators, and equipment that control the physical environment around the system, such as the shift controller in an autonomous vehicle. Table 1 summarizes the common attack surfaces and types for each component of the CPS ecosystem, and possible data sources for forensic investigators. The technical approach for each data source can fall under a different scope of forensics as shown in Figure 2.

The **sensors and actuators**, unfortunately, collect and produce vast amounts of data to be processed in a short time, but they do not store or log information meticulously. The sensor and actuator data are usually not persistent and mostly not accessible to the investigator. Therefore, it is challenging to investigate the data input after an attack if there is no conflicting information from other sensors and if this conflict is not logged.

Some CPS mechanisms utilize sophisticated algorithms or ML/AI mechanisms to implement a **decision-making component** that may or may not be controlled by the main processing unit controller. These decision-making mechanisms require specialized knowledge and logs to explain the decisions made by black-box ML mechanisms [29,30].

The **physical controllers** are also subject to cyberattacks, and depending on the nature of the controller, the approach of the investigation significantly varies. Some controllers can possess their own processing power and the forensic approach can align closer to the processing unit controller. However, another vector of attack on physical controllers is a physical attack that can aim to vandalize, destroy, or decapitate the unit. In these cases, the forensic approach will

Table 1. Common attack types and surfaces of the CPS ecosystem and forensic data source.

Device/ Component	Common Attack Surface Types	Common Forensic Sources
Processing Unit Controllers	• Keyboard • Mouse • I/O devices • Software vulnerabilities • Hardware vulnerabilities • Malicious software	• Application logs • System logs • Security logs • Event managers • Configuration • Workload • OS resources • Storage devices • Known vulnerabilities
Sensors and Actuators	• Input manipulation • Deactivation • By passing	• Physical inspection • Logs
Physical Controllers	• Vandalism • Decapacitation • Destroying • Active manipulation	• Physical inspection • Logs
Network Equipment	• Communication protocol • Flooding • Limited processing power • Spoofing • Sniffing	• Device logs • Known vulnerabilities
Cloud Infrastructure	• Insider threats • Network activity • Connectivity • Attacks to the cloud provider	• Provider logs • Known vulnerabilities
AI/ML Decision Mechanism	• Adaptive features • Adversarial inputs • Data poisoning • Model stealing	• Decision logs • Changes in model over time • Output performance changes
Human	• Phishing and scams • Device jacking • Human error	• Interviews • Device inspection • DNA analysis

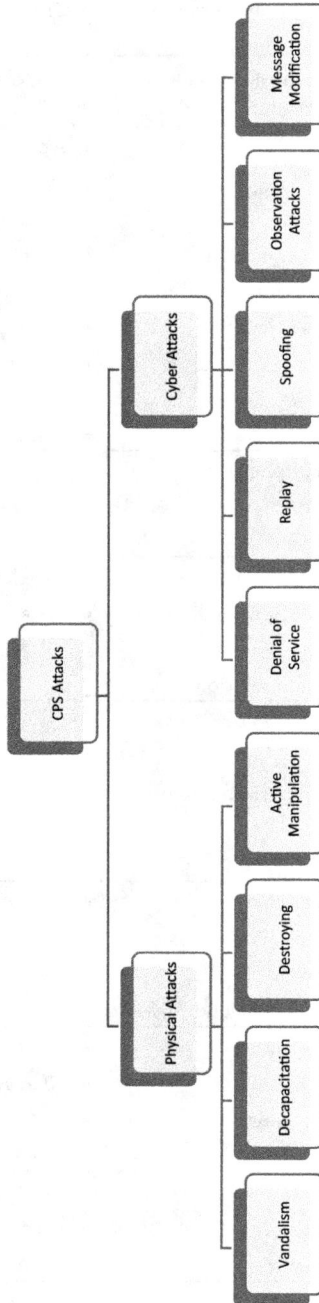

Figure 2. Physical and active CPS attacks.

require DNA analysis, examining crime scene devices and content, and crime scene simulations [31].

The **processing unit controllers** — **computers** — can be considered a traditional computing system that stores a variety of information on the decision, connectivity, and operation of the system. While they are subject to cyberattacks, the forensic investigation approach does not change significantly from traditional digital devices. The **network equipment** that connects the components of the system unfortunately uses mostly volatile data that are usually not persistent and accessible to the investigator. Network components can keep their own logs, but these logs usually expire after a certain time with a first-in-first-out mentality. Some CPS implementations may include special equipment or software to keep the connectivity activity logs, which can be the main tool to investigate incidents.

Lastly, **cloud infrastructure** may also be a part of the CPS in question. The cloud enables users to access the same files, applications, and computing resources from almost any device, because the computing and storage take place on servers in a data center, instead of locally on the user's device. The biggest challenge of performing any assessment on the cloud is that the cloud infrastructure usually does not belong to the CPS owner; therefore, legal processes must take place to access essential investigative data that may be stored in the cloud.

4. Investigation Process

Digital forensics is the science and craft of reconstructing an event or crime, and collecting evidence to build a case to prove the occurrence of a certain action [32].

There are a number of digital forensic investigation frameworks proposed in the literature, and most agree that the fundamental principles of digital forensic investigations are **Reconnaissance**, **Reliability**, and **Relevancy** [33]. In connection with these principles, a forensic investigator must always ask the following seven questions for **each event** that we will refer to as *action survey*:

- Who — the people and the actors
- What — data attributes to be extracted
- Why — the motivation of the action
- How — the procedure to be followed
- Where – the location of the action
- When — the time of the action
- Scope — active device(s)

These questions can incorporate the investigators, legal advisors, and prosecutors into a bigger picture of the investigation, while also exploring the opportunities for reconnaissance, evaluating the reliability of the information and method, and establishing the relevancy of the event to the investigation.

CPS forensics requires investigation of a series of devices for each event, rather than inspection of a single device. Therefore, an investigator must assess the nature of the event to determine the proper technical approach for each device category as shown in Figure 2. CPS poses many challenges for forensic investigators. The variety of information, unclear lines of differentiation between networks and connecting systems, and involvement of cloud mechanisms and sensors complicate the process of the investigation and the structure of the action surveys for each event [34]. However, we can adapt the forensic investigation frameworks [35], as shown in Figure 3, to CPS. This framework performs the technical investigation in 5 steps: (i) preparation, (ii) evidence collection and preservation, (iii) examination and analysis, (iv) presentation and reporting, and (v) disseminating the case.

In Table 2, we provide details of each step of the technical investigation with generalized activities to be performed.

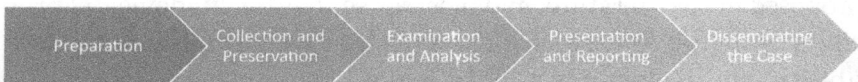

Figure 3. Step-by-step framework.

Table 2. Step-by-step outline of the technical investigation.

Step	Activity	Outcomes
Preparation	• Establish if there is a need for investigation • Establish probable cause for the investigation • Identify potential data sources • Prepare equipment to support the investigation • Inform concerned parties of the investigation • If the investigation concerns private property or data, send preservation orders and request necessary warrants • Plan for the next steps	• Collection plan • Preservation plan • Analysis plan • Action items • Authorization
Collection and Preservation	• Make sure to have proper authority to access a potential data source • Confirm that the data source is viable • Record the preparation and collection process • Use write blockers to access any digital storage • Make examinable copies of the original while making sure that the evidence is admissible • Package and store the evidence properly based on its nature • Log every interaction with the data source and evidence	• Evidence sources
Examination and Analysis	• Assess the skill level of the suspect against anti-forensic techniques • Identify the methods to extract evidence • Validate the efficacy of the selected methods • Extract obvious pieces of digital evidence	• Event log • Supplementary information • Branching of the investigation

(Continued)

Table 2. (*Continued*)

Step	Activity	Outcomes
	• Discover and extract hidden digital evidence	
	• Cross-match the personal data with the device and application logs, if applicable	
	• Analyze the data using the action survey	
	• Cross-validate potential events with evidence	
	• Construct, test, and reject alternative timelines	
	• Document the findings	
Presentation and Reporting	• Clarify the evidence and supplement findings	• Analyzed and cataloged evidence
	• Prove the validity of evidence against challenge	• Report for a variety of audiences
	• Find impartial expert witnesses, if applicable	
	• Construct an explanation of conclusions for each viable hypothesis of timeline(s)	
	• Prepare reports for law enforcement, management, and public with proper language	
Disseminating the Case	• Review the investigation process to find errors and improvement areas	• Explanation of evidence and timeline
	• Return the digital property to proper owner	• New procedures
	• Distribute the report to concerned parties	• New policies

5. Forensic Approaches and Analysis

5.1. *Legal perspective*

While the collection of digital evidence requires the technical expertise and tools necessary to curate information as evidence, there are legal factors that forensic investigators must consider and act on, such as *personal privacy* or even human rights. In the United States, law enforcement must obtain a warrant issued by a judge or magistrate before a search or arrest can be carried out.

A **search warrant** is a court order issued by a judge or magistrate authorizing law enforcement to search a person or place, as well as seize items or information within the parameters of the warrant. Moreover, an investigator must demonstrate **probable cause**, which refers to the conditions under which law enforcement may obtain a warrant for a search or arrest when it is evident that a crime has been committed.

It is important to note that the process of acquiring a warrant, its scope, the definition of evidence, and the rules of evidence admissibility may vary based on the state or country.

Acquiring a warrant may also take time and effort in certain cases. To prevent data and potential evidence loss, there usually is a mechanism commonly called a *preservation order*. The *preservation order* is relatively easy to issue in many jurisdictions and aims to preserve case-related data that may or may not contain relevant evidence for a certain time. It may also be called a litigation hold or a hold order. It is mostly sent directly to a party, instructing the recipient not to destroy, alter, or delete any documents helpful to the sender.

5.2. *Technical perspective*

To achieve reliable operation in a CPS environment, it is crucial to accurately create a threat model that defines the attacker profile, possible motives, and the system components that they can target. This evaluation can lead the investigators to the accurate data sources to extract meaningful evidence.

5.2.1. *Sensors and actuators analysis*

The sensor and actuator infrastructure can comprise various components such as cameras, Light Detection and Ranging (LiDAR), radar, magnetic field sensors, GPS sensors, and RFID sensors.

The sensor infrastructure provides data for the operation of the whole system through AI/ML systems or simpler input-based systems; therefore, tampering of the infrastructure can lead to accidents and malfunction. There have been several instances in the past where attackers were successfully able to fool the sensor systems and control the system they targeted [36,37].

Logging the inputs the sensors send to the system can provide investigators with valuable information. There are three types of forensically relevant data the sensors and actuators can produce: (i) sensory data provided to the system, (ii) failure and malfunction logs, and (iii) warnings on the reliability or dependability of the data provided.

Sensory data provided to the system: Unfortunately, the sensor infrastructure does not usually have a capable logging mechanism to report on sensory data. Some sensors create TBs of data per day, and it is impossible to log at the speed they create data. In the case that the investigator does not have direct data from the sensor or actuator, they can refer to the decision-making mechanism and its faults to extrapolate sensor data.

Failure and malfunction logs: These types of logs are fed into the processing unit controllers. They are the most reliable and consistent investigative data source for the investigator. However, the nonexistence of a certain failure log record must not be interpreted as proof that the sensor or actuator did not malfunction.

Warning on the reliability or dependability: Some sensors are capable of sensing that their input data are not consistent or reliable. For example, on a camera, dirty lenses distort the view, and some cameras are capable of detecting that the quality of the input they provide is low. Similar to the failure and malfunction logs, the nonexistence of this type of log record must not be interpreted as proof that the sensor or actuator provides high-quality data. In fact, most sensors are not capable of sensing the input quality.

5.2.2. *Processing unit/controller analysis*

Processing unit/controller is mostly a commodity or specialized computer. If this computer also carries traditional operating systems (OS) such as Microsoft Windows®, or MacOS®, traditional investigative techniques apply. The investigator should act based on the type of attack or failure, because if the attack only targets the sensor systems or the decision-making mechanism, then the investigator should focus on extracting evidence from the logs and data related to that system from this computer. However, if the attack concerns the confidentiality, integrity, or availability of the processing unit, then the investigator should focus on the OS logs, application logs, event manager data, files, and databases. The investigation will likely require copying of the storage device as well. In this case, the storage device will need to be copied with a lawfully approved cloning device. The jurisdiction may also have requirements that the copy drive be wiped with a certain algorithm such as the DoD 7-Pass Erase mechanism before the data are copied on that device.

5.2.3. *Physical controller analysis*

Physical controllers are the devices that reflect the decisions made by the system to the physical world. These devices can range from simple non-interactive LED screens to robotic precision surgery machines. While the capabilities vary, the most common attacks on these components are physical and manifest themselves in (i) feedback to the decision mechanism, (ii) failure and malfunction logs, and (iii) warnings on the reliability or dependability of the data provided.

Feedback to the system: The feedback infrastructure does not usually have a capable logging mechanism to report on feedback data. Similar to the sensors, it may create TBs of data per day or activity, and it is impossible to log at the speed it creates data. In the case that the investigator does not have direct feedback records, they can refer to the decision-making mechanism, their faults, or to the human operator to extrapolate the feedback.

Failure and malfunction logs: Similar to sensor data, these types of logs are fed into the processing unit controllers. They are the most reliable and consistent investigative data source for the investigator. However, nonexistence of a certain failure log record must not be interpreted as proof that the sensor or actuator did not malfunction.

Warning on the reliability or dependability: Similar to sensor data, some control devices are capable of sensing that their input data are not consistent or reliable. For example, autonomous vehicle (AV) wheels may turn without any friction, which can mean that the car is spinning that wheel on the ground or another unlikely surface. The system may need additional data from other systems to assess the situation. Similar to the failure and malfunction logs, nonexistence of this type of log record must not be interpreted as proof that the physical component provides high-quality data. In fact, most physical controllers are not capable of sensing the input quality.

The physical controllers are controlled by the processing unit and mostly depend on software dedicated to them as a part of the main system, or as parameter-based independent software. In this case, the logs generated by the software must also be considered for investigation.

5.2.4. *Network equipment analysis*

In most cases, the network equipment uses standard protocols for communication with the cloud or internally in the system. Like all networked systems, CPSs are vulnerable to attacks against their confidentiality, integrity, and availability. The investigative techniques for all networked systems apply to CPS as well. However, the most common components of the system to be inspected are network/router logs, packet transfers, firewall logs, OS logs, and application logs.

5.2.5. *Cloud infrastructure analysis*

Some CPSs are also connected to a cloud infrastructure. The role of the cloud may widely vary based on the nature of the CPS.

For example, the cloud can also be the processing unit for the data and, in some cases, it may only be used for data storage or backup. Based on this relationship, the CPS may be vulnerable to all the cloud system's vulnerabilities.

Another aspect that the inspector must consider is the fact that the cloud infrastructure and the CPS are not operated or owned by the same parties. Due to this, the common practice is to identify any cloud-related components and issue *preservation orders* to the cloud providers related to the CPS to prevent any possible data loss until the warrants to acquire relevant data come through.

5.2.6. *Decision-making component analysis*

Sensor infrastructures comprise three phases, i.e., *Sense, Understand,* and *Action.* After collecting raw sensor data from its surroundings using multiple sensors, a CPS generates an image of certain relevant fragments of the environment by fusing the data to the *Action Engine.* While making incorrect judgments is always possible for an ML mechanism, feeding bad or fake data from sensors can lead to improper decisions. It is desirable to distinguish between simple decision mistakes and attacks. ML methods, specifically *deep neural networks* used by CPS, process sensor inputs by sending them through a network of nodes analogous to neurons. Sensor inputs pass from node to node along colliding links as the synaptic junctions between neurons. Each junction is named a *hidden layer.* The success of the *learning* improves by adjusting the weights that amplify or damp the value of the link for specific inputs.

Most CPSs follow the OODA (*observe–orient–decide–act*) loop [38]. Therefore, it is necessary to document the events that lead to each decision that the CPSs make based on their OODA loop prior to a failure from the logs. The first challenge is to identify, capture, and label the input parameters of the environment.

ML/AI methods have been widely used in vehicle technologies such as V2V communication [39], behavior planning [40], and driving [41]. However, the explainability of AI has recently become a topic of interest [42–44], and researchers and practitioners have been

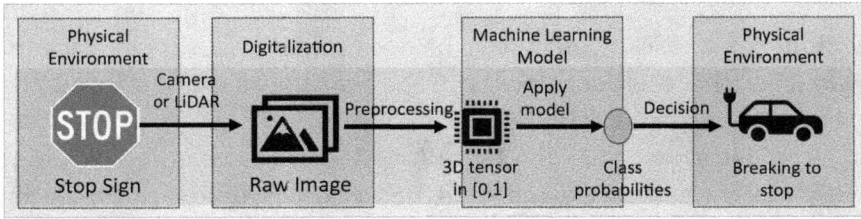

Figure 4.   Decision-making process.

calling for wider support for the development of explainable AI (XAI) techniques [45]. Generally, the application areas are in fairness in AI and ML [43], cognitive engines [46], and the medical field [47]. With XAI, for each decision the control mechanism is making, we should be able to create a flowchart of the decision made and extract the position of the neuron that leads to a faulty action.

For example, in an AV, a **sensor-jamming** attack uses the attack surface of the physical environment and corrupts the sensor data quality. The attacker focuses on provoking an accident, disrupting road traffic, risky lane changing, and/or controlling the car remotely [48]. Such an action can corrupt a stop sign. Normally, the decision-making mechanism feeds the digital representation of the information to the deep learning model and decides to apply breaks to stop at the sign as shown in Figure 4. If the sign data are corrupted or manipulated through an attack, our forensic model will be able to identify the *sign* neuron and track the events that led to the accident.

6.   Conclusion

In the first half of 2021, hacker groups successfully conducted cyberattacks on critical infrastructure, pipelines, meat manufacturers, and state government systems. Attacks against CPS using the vulnerabilities in the commonly used technologies are already common. Since CPSs are built using a variety of components, they inherit their vulnerabilities as well. Hackers and hacker groups use this as a weapon to threaten, disrupt, and interrupt people, governments, and organizations that depend on these systems. CPS owners and operators are usually neither equipped nor ready to forensically investigate malfunctions, attacks, or failures concerning the cyber realm. The primary duty of the operators is usually to provide safety to the people, and then bring the system back to life to minimize downtime. Therefore, forensic investigators must be prepared to act independently and with minimal help from the owners or operators even though they act for the benefit of the same parties.

References

[1] Derler, P., Lee, E.A., and Vincentelli, A.S. Modeling cyber–physical systems. *Proceedings of the IEEE*, 100(1):13–28, 2011.

[2] Radanliev, P., De Roure, D., Van Kleek, M., Santos, O., and Ani, U. Artificial intelligence in cyber physical systems. *AI & Society*, 36(3):783–796, 2021.

[3] Lee, E.A. Cyber physical systems: Design challenges. In *2008 11th IEEE International Symposium on Object and Component-Oriented Real-Time Distributed Computing (ISORC)*, pp. 363–369, 2008.

[4] Caviglione, L., Wendzel, S., and Mazurczyk, W. The future of digital forensics: Challenges and the road ahead. *IEEE Security Privacy*, 15(6):12–17, 2017.

[5] Chattopadhyay, A., Prakash, A., and Shafique, M. Secure cyber-physical systems: Current trends, tools and open research problems. In *Design, Automation & Test in Europe Conference & Exhibition (DATE)*, pp. 1104–1109. IEEE, 2017.

[6] Garfinkel, S.L. Digital forensics research: The next 10 years. *Digital Investigation*, 7:S64–S73. 2010, the *Proceedings of the Tenth Annual DFRWS Conference*. Available at: https://www.sciencedirect.com/science/article/pii/S1742287610000368.

[7] Sonntag, D., Zillner, S., Chakraborty, S., Lorincz, A., Strommer, E., and Serafini, L. The medical cyber-physical systems activity at EIT: A look under the hood. In *2014 IEEE 27th International Symposium on Computer-Based Medical Systems*, pp. 351–356. IEEE, 2014.

[8] Hussain, I. and Park, S.J. Big-ECG: Cardiographic predictive cyber-physical system for stroke management. *IEEE Access*, 9:123146–123164, 2021.

[9] Li, Y.-T., Jacob, M., Akingba, G., and Wachs, J.P. A cyber-physical management system for delivering and monitoring surgical instruments in the OR. *Surgical Innovation*, 20(4):377–384, 2013.

[10] Klimeš, J. Using formal concept analysis for control in cyber-physical systems. *Procedia Engineering*, 69:1518–1522, 2014.

[11] Medhat, R., Bonakdarpour, B., Kumar, D., and Fischmeister, S. Runtime monitoring of cyber-physical systems under timing and memory constraints. *ACM Transactions on Embedded Computing Systems (TECS)*, 14(4):1–29, 2015.

[12] Han, R., Zhao, X., Yu, Y., Guan, Q., Hu, W., and Li, M. A cyber-physical system for girder hoisting monitoring based on smartphones. *Sensors*, 16(7):1048, 2016.

[13] Duan, Y., Fu, X., Luo, B., Wang, Z., Shi, J., and Du, X. Detective: Automatically identify and analyze malware processes in forensic scenarios via DLLs. In *2015 IEEE International Conference on Communications (ICC)*, pp. 5691–5696. IEEE, 2015.

[14] Li, J., Gu, D., and Luo, Y. Android malware forensics: Reconstruction of malicious events. In *2012 32nd International Conference on Distributed Computing Systems Workshops*, pp. 552–558. IEEE, 2012.

[15] Thing, V.L., Ng, K.-Y., and Chang, E.-C. Live memory forensics of mobile phones. *Digital Investigation*, 7:S74–S82, 2010.

[16] Graziano, M., Lanzi, A., and Balzarotti, D. Hypervisor memory forensics. In *International Workshop on Recent Advances in Intrusion Detection*, pp. 21–40. Springer, 2013.

[17] Wang, W. and Daniels, T.E. A graph based approach toward network forensics analysis. *ACM Transactions on Information and System Security (TISSEC)*, 12(1):1–33, 2008.

[18] Ndatinya, V., Xiao, Z., Manepalli, V.R., Meng, K., and Xiao, Y. Network forensics analysis using wireshark. *International Journal of Security and Networks*, 10(2):91–106, 2015.

[19] Sathe, S.C. and Dongre, N.M. Data acquisition techniques in mobile forensics. In *2018 2nd International Conference on Inventive Systems and Control (ICISC)*, pp. 280–286. IEEE, 2018.

[20] Mahalik, H., Tamma, R., and Bommisetty, S. *Practical Mobile Forensics*. Packt Publishing Ltd, Birmingham, UK, 2016.

[21] Khanuja, H.K. and Adane, D. Detection of suspicious transactions with database forensics and theory of evidence. In *International Symposium on Security in Computing and Communication*, pp. 419–430. Springer, 2018.

[22] Nemetz, S., Schmitt, S., and Freiling, F. A standardized corpus for SQLite database forensics. *Digital Investigation*, 24:S121–S130, 2018.

[23] Abedi, M. and Sedaghat, S. Crawler and spiderin usage in cyber-physical systems forensics. *OIC-CERT Journal of Cyber Security*, 1(1):53–61, 2018.

[24] Chan, R. and Chow, K.-P. Forensic analysis of a siemens programmable logic controller. In *International Conference on Critical Infrastructure Protection*, pp. 117–130. Springer, 2016.

[25] Mishra, S. Forensic investigation framework for complex cyber attack on cyber physical system by using goals/sub-goals of an attack and epidemics of malware in a system. In *Recent Trends in Communication, Computing, and Electronics*, pp. 491–504. Springer, 2019.

[26] Alrimawi, F., Pasquale, L., Mehta, D., and Nuseibeh, B. I've seen this before: Sharing cyber-physical incident knowledge. In *Proceedings of the*

1st International Workshop on Security Awareness from Design to Deployment, pp. 33–40, 2018.

[27] Al-Sharif, Z.A., Al-Saleh, M.I., Alawneh, L.M., Jararweh, Y.I., and Gupta, B. Live forensics of software attacks on cyber–physical systems. *Future Generation Computer Systems*, 108:1217–1229, 2020.

[28] Kul, G., Upadhyaya, S., and Hughes, A. An analysis of complexity of insider attacks to databases. *ACM Transactions on Management Information Systems*, 12(1):2020 [Online]. Available at: https://doi.org/10.1145/3391231.

[29] Sharma, P., Liu, H., Wang, H., Zhang, S. Securing wireless communications of connected vehicles with artificial intelligence. In *Proceedings of the 2017 IEEE International Symposium on Technologies for Homeland Security (HST)*, Waltham, MA, 2017, pp. 1–7, doi: 10.1109/THS.2017.7943477.

[30] Sharma, P., Siddanagaiah, U., and Kul, G. Towards an AI-based after-collision forensic analysis protocol for autonomous vehicles. In *2020 IEEE Security and Privacy Workshops (SPW)*, pp. 240–243. IEEE, 2020.

[31] Mohamed, N., Al-Jaroodi, J., and Jawhar, I. Cyber–physical systems forensics: Today and tomorrow. *Journal of Sensor and Actuator Networks*, 9(3):37, 2020.

[32] Casey, E. *Handbook of Digital Forensics and Investigation*. Academic Press, Burlington, MA, 2009.

[33] Ieong, R.S. FORZA — digital forensics investigation framework that incorporate legal issues. *Digital Investigation*, 3:29–36, 2006. *Proceedings of the 6th Annual Digital Forensic Research Workshop (DFRWS'06)*. Available at: https://www.sciencedirect.com/science/article/pii/S1742287606000661.

[34] Sathwara, S., Dutta, N., and Pricop, E. IoT forensic a digital investigation framework for IoT systems. In *2018 10th International Conference on Electronics, Computers and Artificial Intelligence (ECAI)*, pp. 1–4, 2018.

[35] Selamat, S.R., Yusof, R., and Sahib, S. Mapping process of digital forensic investigation framework. *International Journal of Computer Science and Network Security*, 8(10):163–169, 2008.

[36] Petit, J. and Shladover, S.E. Potential cyberattacks on automated vehicles. *IEEE Transactions on Intelligent Transportation Systems*, 16(2):546–556, 2014.

[37] Petit, J., Stottelaar, B., Feiri, M., Kargl, F. Remote attacks on automated vehicles sensors: Experiments on camera and LIDAR. In *Proceedings of the 2015 Black Hat Europe*, Amsterdam, The Netherlands, 2015, p. 995.

[38] Huang, H.-M., Pavek, K., Albus, J., and Messina, E. Autonomy levels for unmanned systems (ALFUS) framework: An update. In *Unmanned Ground Vehicle Technology VII*, 5804, pp. 439–448. International Society for Optics and Photonics, 2005.

[39] Huang, C., Molisch, A.F., He, R., Wang, R., Tang, P., and Zhong, Z. Machine-learning-based data processing techniques for vehicle-to-vehicle channel modeling. *IEEE Communications Magazine*, 57(11):109–115, 2019.

[40] Sun, L., Zhan, W., Chan, C.-Y., and Tomizuka, M. Behavior planning of autonomous cars with social perception. In *2019 IEEE Intelligent Vehicles Symposium (IV)*, pp. 207–213. IEEE, 2019.

[41] Tian, Y., Pei, K., Jana, S., and Ray, B. Deeptest: Automated testing of deep-neural-network- driven autonomous cars. In *Proceedings of the 40th International Conference on Software Engineering*, pp. 303–314, 2018.

[42] Gunning, D. and D. Aha. DARPA's Explainable Artificial Intelligence (XAI) Program. *AI Magazine*, vol. 40, no. 2, June 2019, pp. 44–58, doi:10.1609/aimag.v40i2.2850.

[43] Adadi, A. and Berrada, M. Peeking inside the black-box: A survey on explainable artificial intelligence (XAI). *IEEE Access*, 6:52138–52160, 2018.

[44] Došilović, F.K., Brčić, M., and Hlupić, N. Explainable artificial intelligence: A survey. In *2018 41st International Convention on Information and Communication Technology, Electronics and Microelectronics (MIPRO)*, pp. 0210–0215, 2018.

[45] Samek, W. and Müller, K.-R. *Towards Explainable Artificial Intelligence*, pp. 5–22. Springer International Publishing, Cham, 2019.

[46] Saghiri, A.M. A survey on challenges in designing cognitive engines. In *2020 6th International Conference on Web Research (ICWR)*, pp. 165–171. IEEE, 2020.

[47] Watson, M. and Al Moubayed, N. Attack-agnostic adversarial detection on medical data using explainable machine learning. In *2020 25th International Conference on Pattern Recognition (ICPR)*, pp. 8180–8187. IEEE, 2021.

[48] Cao, Y., Xiao, C., Cyr, B., Zhou, Y., Park, W., Rampazzi, S., Mao, Z. M. Adversarial sensor attack on lidar-based perception in autonomous driving. In *Proceedings of the 2019 ACM SIGSAC Conference on Computer and Communications Security (ACM CCS)*, pp. 2267–2281. 2019.

Part 2
Applications

© 2024 World Scientific Publishing Company
https://doi.org/10.1142/9789811273551_0004

Chapter 4

Cloud Forensic: Issues, Challenges, and Solution Models

Sayada Sonia Akter* and Mohammad Shahriar Rahman†

*Department of Computer Science & Engineering,
United International University, United City,
Madani Avenue, Badda, Dhaka 1212, Bangladesh*
*sakter212039@mscse.uiu.ac.bd
†mshahriar@cse.uiu.ac.bd

Abstract: Cloud computing is a web-based utility model that is becoming popular every day with the emergence of 4th Industrial Revolution; therefore, cybercrimes that affect web-based systems are also relevant to cloud computing. In order to conduct a forensic investigation into a cyberattack, it is necessary to identify and locate the source of the attack as soon as possible. Although significant analysis has been done in this domain on obstacles and their solutions, research on approaches and strategies is still in the development stage. There are barriers at every stage of cloud forensics; therefore, before we can come up with a comprehensive way to deal with these problems, we must first comprehend cloud technology and its forensic environment. Although there are articles that are linked to cloud forensics, there is no paper as yet on the accumulated contemporary concerns and solutions related to cloud forensics. Throughout this chapter, we have looked at the cloud environment, as well as the threats and attacks that it may be subjected to. We have also looked at the approaches that cloud forensics may take, as well as the various frameworks, the practical challenges, and limitations one may face when dealing with cloud forensic investigations.

Keywords: Cloud Environment, Threats, Cloud Forensic, Frameworks.

1. Introduction

Cloud Computing is increasingly becoming one of the most advanced technologies, with great future prospects for companies and organizations in the coming years. This era has experienced a revolution in cloud computing which not only caused many to consider the concept as a new Information Technology architecture but also gave it a reputation of being the most fast-changing and industry-altering technology since the invention of computers [1]. In 2020, cloud computing services produced more than $300 billion in sales. Considering the most recent cloud computing data, it is a fair bet that this business will only grow in the next decade [3]. Moreover, this massive growth has transformed how IT might be used to access, administer, develop, and offer services [2]. Indeed, using cloud computing in enterprises may lower the cost of IT [3], which makes us feel that money is one of the key reasons why cloud computing is considered a quickly expanding technology.

However, the rapid and extensive use of cloud computing has created a situation where cloud environments are currently seen as a new potential environment for the execution of cybercriminal activities. This circumstance has also resulted in the emergence of completely new legal, organizational, and technological obstacles. At this moment, it is relevant to mention the considerable number of attacks that have an influence on cloud computing as well as the fact that cloud-based data processing is carried out in a manner that is decentralized; in fact, in addition to these factors, many people have voiced concerns about how a comprehensive digital investigation can be undertaken in environments that are hosted by cloud providers [1]. Generally, it is vital to undertake investigations independently, without the need to depend on a third-party provider. But, things are different in virtualized environment, where this process remains challenging. These complications stem from the fact that cloud services, who fully determine what takes place in the environment, keep control of the sources of evidence. Consequently, it is still hard to determine who is responsible for what. In addition, customers are still unable, at least to some degree, to proactively acquire data ahead of the occurrence of an event [4]. Thus, ensuring that forensic readiness has been achieved prior to conducting digital investigations will lower the amount of money as well as time that would be required on the investigations. As per Market Research Media [14], the worldwide cloud computing industry has been forecasted to grow by a compound annual growth (CAGR) of 30% though the 2025; at this point, many feel that the market will be worth nearly $270 billion. This figure effectively demonstrates the expansion of the cloud computing sector as well as the rapid growing number of users in the cloud around the world. This expansion will directly result in an increase in the number of cyberattacks. Although cloud forensics has a significant number of challenges, there are currently no regulations, processes, or standards established [5].

This chapter's objective is to provide the readers with an overview of Cloud Environment as well as the security challenges related to it. We also provide an analysis of a few prominent cloud-based cyberattacks. This sheds light on the need for cloud-based forensics as well as on the challenges associated with it. The National Institute of Standards and Technology (NIST) recently compiled a list of sixty-five cloud forensic challenges [6]. We go further to identify potential solutions to these challenges by analyzing different cloud-based forensic models as suggested by different authors. Finally, this research study may be regarded as an endeavor to encourage continuing research in this sector and build cloud forensic-capable systems.

The readers were given an introduction to cloud computing in the first portion of this chapter. The second section focuses more on cloud environment, including its characteristics and the standards that govern them. In the third section, we evaluate a few prominent cloud-based cyberattacks and offer our observations on the reasons why these events occurred, which makes it clearly evident why it is essential to have cloud forensics in order to be forensically ready. The readers will have a fundamental understanding of cloud forensics after reading the fourth and sixth sections. In the seventh section, readers will get a grasp of Cloud Forensic Challenges. In the eighth section, we will evaluate numerous models to address a number of significant challenges listed by NIST. In section nine, we offer an overview of the cloud forensic tools that are now accessible. Last but not least, we bring our review chapter to a close by including sections ten, eleven, and twelve that discuss potential opportunities and directions for future research.

2. Cloud Computing Environment

Cloud computing, as described by the National Institute of Standards and Technology [NIST], has "revolutionized the techniques through which digital data is stored, processed, and transported. It enables ubiquitous, accessible, on-demand network access to a shared pool of customizable computer resources (e.g., networks, servers, storage, applications, and services) that can be instantly supplied and released with minimum administrative effort or service provider contact" [6].

On-Demand Computing, or "Cloud Computing", is a kind of computing that relies solely on the Internet. Computers and other electronic devices may now be supplied with data, information, and other necessary shared resources on demand using this computational technology. It is possible to store and process data in many third-party data centers by providing cloud computing and storage options to numerous customers and businesses. As converged infrastructure and shared services are at the heart of this new technology, this technology separates information resources from the underlying infrastructure and delivery mechanisms.

2.1. *Characteristics of cloud computing*

Sharing resources is essential to cloud computing's goal of achieving coherence while also achieving cost efficiencies. Cloud computing allows businesses to execute their applications in a more efficient manner since resources are made available depending on the needs of the application. This technology was developed to produce systems that are efficient in terms of cost, allowing businesses to operate their processes without encountering any IT roadblocks. In cloud computing, a method known as virtualization was developed to partition a physical device into one or more virtual devices. This allowed efficient use of available resources. Computing in the cloud is comparable to computing on a grid. The usefulness of this technology may be associated with a number of factors, including the following:

- Self-Service on Demand: The cloud service's consumers can use various services on demand without the help of the service providers.
- Pooling Resources: When resources are pooled, users may be served in one location or several locations, based on what is most beneficial for them individually.
- Wide Network Access: Resources related to computing are sent across the network. Consumers can use the service once there is a network connection in their environment.
- Scalability: The resources can be set up without the help of service providers and can be rapidly scaled up or down to fit the requirements of the user.
- Measured Service: The cloud architecture is able to make use of the appropriate strategies despite the fact that the computer resources are pooled and shared across a number of different participants.
- Maintenance: Because of its accessibility, the maintenance of cloud apps is quite straightforward.
- Multi-tenancy: Multi-tenancy is a useful feature that allows for cost and resource sharing.

2.2. *Cloud computing standards*

The international standard ISO/IEC 17788 defines a wide variety of service models for cloud computing. Within the community of cloud computing, three primary standards have been developed – Service, Deployment, and Role [24]. To begin, the cloud computing industry has relied heavily on three service models to classify cloud services, as shown in Figure 1.

S. S. Akter & M. S. Rahman

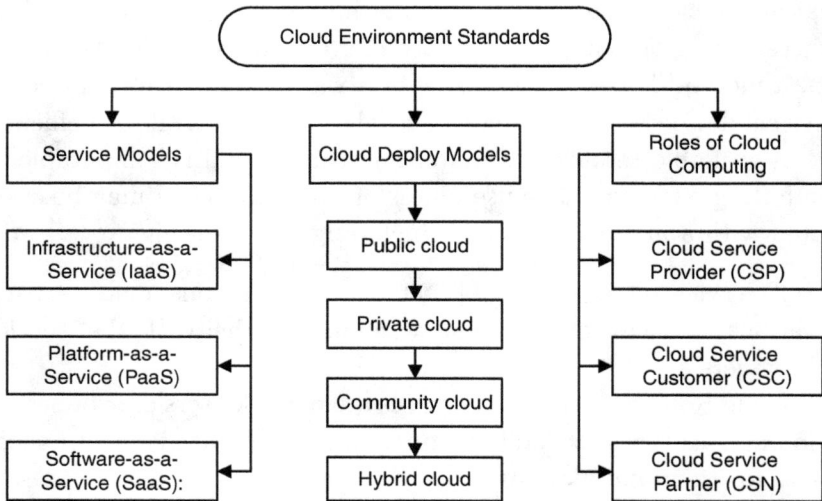

Figure 1.   Cloud environment standards.

- Infrastructure-as-a-Service (IaaS): IaaS is a kind of cloud computing that provides basic computing, network, and storage capabilities to customers on request, through the internet, on a pay-as-you-go basis.
- Platform-as-a-Service (PaaS): When it comes to developing, running, and managing applications, PaaS is a model of cloud computing that provides customers with a complete cloud platform that includes hardware, software, and infrastructure without the cost, complexity, and inflexibility that are typically associated with building and maintaining that platform on-premises.
- Software-as-a-Service (SaaS): The SaaS framework makes it possible to provide software applications to end customers in the form of a service. It is a term used to describe a piece of software that has been installed on a host service and is available to users through the internet.

Cloud computing can be broken down into four categories: public, private, community, and hybrid:

- Public cloud: Cloud services are accessible to every cloud service client, and the cloud service provider is in charge of the resources.

- Private cloud: Cloud services are used by a specific cloud service client, who also retains full control over the resources made available via such cloud services.
- Community cloud: A cloud environment that is shared among several entities and belongs to the same community is referred to as a "community cloud". It is the cloud infrastructure environment that provides the same level of privacy and security as a private cloud.
- Hybrid cloud: A hybrid cloud, also known as a cloud hybrid, is a type of cloud computing system that combines an on-premises data center, also known as a private cloud, with a public cloud. This makes it possible for users' data and applications to be shared between these two types of clouds.

At last, three distinct functions for cloud computing have been outlined: Cloud Service Provider (CSP), Cloud Service Customer (CSC), and Cloud Service Partner (CSN).

- Cloud Service Provider (CSP): A third-party organization that offers platform, infrastructure, application, or storage services through the cloud is known as a CSP.
- Cloud Service Customer (CSC): A subscriber to a cloud service and a user of a cloud service are both examples of CSCs. Within the limitations of a private cloud, all of the users and the service providers belong to the same company.
- Cloud Service Partner (CSN): A partner that allows people to access the public cloud services via their own interface is known as a CSN. They are also responsible for billing and providing administrative assistance for billing-related matters.

2.3. *Threats in cloud environments*

Users of cloud computing do not know the precise location of their sensitive data since Cloud Service Providers (CSPs) operate their data centers in a variety of locations throughout the world; as a result, users are put in a vulnerable position regarding their safety. Standard security measures such as host-based antivirus software,

Table 1.   CSA'S Top 12 threats.

Threat No.	Threat Name
1	Data breaches
2	Compromised credentials and broken authentication
3	Hacked interface and Application Program Interfaces
4	Exploited system vulnerabilities
5	Account hijacking
6	Malicious insiders
7	The Advanced Persistent Threat (APT) parasite
8	Permanent data loss
9	Inadequate diligence
10	Cloud service abuses
11	Denial-of-Service (DoS) attacks
12	Shared technology, shared dangers

intrusion detection systems, and firewalls do not offer enough protection for virtualized systems because of the rapid spread of threats that occur in virtualized settings. According to Walker [25], the Cloud Security Alliance (CSA) has compiled and released a list of the top 12 threats that have been found to be associated with cloud computing. These potential issues are stated in Table 1. Out of all these potential dangers, the compromising of data has been pointed out as the most significant security problem that has to be addressed.

Several authors [26–32] came to the conclusion that the difficulties highlighted in Table 2 are preventing solutions from being established for threats. They consider these challenges to be the gaps for threat remediation and feel that they must be addressed in further study. It is interesting to note that a lot of these problems are trust concerns, which may result in cyberattacks. These difficulties may be addressed from both directions: Cloud service providers do not trust their consumers and believe that their customers may bypass existing privacy rules in order to carry out phishing and malware attacks while using the cloud services that the cloud providers offer [33]. On the other hand, clients who use the cloud have a variety of additional trust difficulties. For instance, service providers may constantly attempt to disguise their own corporate rules for recruitment, and the incapacity of service providers to monitor their employees may result in an

Table 2. Defects in the process of threat remediation.

Security Threats	Challenges While Implementing Threat Remediation
Misuse and Malicious Application of Cloud Computing	• Because of privacy restrictions, cloud service providers are unable to provide real-time monitoring. • The interests of many stakeholders do not necessarily align in the same manner.
Insecure Application Programming Interfaces	• Inability to audit events connected to the usage of APIs. • Incomplete data from the logs to allow the reconstruction of company's strategies.
Malicious Insiders	• Service providers may sometimes attempt to conceal their own corporate policies while hiring company employees. • The solutions are implemented after the occurrence has already taken place, which is too late. • The incapacity of cloud service providers to monitor their staff members.
Vulnerabilities Existing in Shared Technology	• Elements that are shared have never been intended for a high level of fragmentation. • Competitors in the business world use distinct virtual machines hosted on the same physical machine. • The presence of both a retail and a manufacturing sector together.
Loss/Leakage of Confidential Data	• Lack of trust in the cloud providers since they could act in their own self-interest and keep data in a less secure location than was agreed upon. • Procedures that have not been thoroughly evaluated, weak policies, and practices for the preservation of data. • A poor understanding.
Hijacking of Accounts, Services, and Traffic	• Faster evolution of cloud computing results in the creation of new security vulnerabilities. • Current method of managing digital identities is insufficient for use with hybrid cloud environments.
Unknown Risk Factors	• The reluctance of cloud service providers to give log and audit data, and procedures regarding cloud security. • Lack of honesty and transparency.

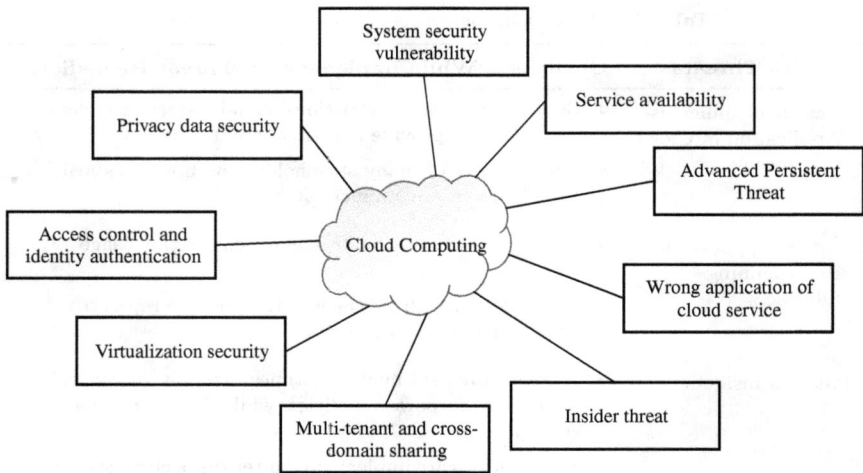

Figure 2.   Privacy security risk in cloud computing.

attack by malicious insiders. Images obtained from unreliable sources might leave a back door open for attackers to use. It is possible for cloud providers to become self-interested and greedy, and as a result, they may employ storage space with a lesser security level than was agreed upon, which may result in Cross VM Side channel attacks.

On the other hand, based on the findings of research conducted by relevant academics and the annual report of the Cloud Security Alliance (CSA) [7], there are significant dangers to the privacy security risk as shown in Figure 2.

- *Identity authentication and access control*: Cloud computing requires an enormous number of resources, which significantly increases the administrative difficulty of managing access controls and identity verification.
- *Privacy data security*: As a result of the method of service outsourcing, the security risk associated with cloud privacy is very prominent. This risk includes challenges with data disclosure, privacy disclosure, access rights management, and data deletion.
- *Multi-tenant and cross-domain sharing*: Multi-tenant and cross-domain sharing requires the assurance of multi-tenant isolation

and multi-user security. When a domain is crossed, service authorization and access control become more difficult to manage, and the process of trust transfer among two cloud computing organizations needs to be reevaluated.

- *Virtualization security*: Although service providers have designed and implemented isolation techniques for virtual machines, it is impossible to completely avoid attacks between virtual machines. Migration of machines that are virtualized will also produce changes in the security domain. Despite these efforts, attacks between virtual machines can still occur.
- *Vulnerabilities of system security*: Because of the complexity of a cloud computing system, many service providers provide various management and service levels. As a result, the risk associated with using the cloud will grow if there are any security flaws.
- *Advanced persistent threat (APT)*: It refers to an infiltration and attack that have been prepared on a cloud computing system. This system has developed certain underground interest chains.
- *Wrong application of cloud service*: Misusing cloud computing may result in difficulties for customers, service providers, or third parties, and illegally using cloud services will result in severe consequences.
- *Unviability of cloud service*: Denial-of-service attacks have become a significant security target for cloud service providers as a result of the fact that many security incidents are expressed as the inability of cloud computing services to be accessed.
- *Threats generated by insiders*: It is common for the security policy to be ineffective due to the accidental or purposeful disclosure of sensitive information by service provider insiders. This has developed into a major concern for cloud computing security.

3. Taxonomy of Cloud Based Cyber-Attacks

Any cloud-based service can be vulnerable to security breaches due to a number of factors, including misconfiguration, unauthorized access, insecure interfaces or application programming interfaces (APIs), account hijacking, or even malicious insiders. In this section, we will discuss a few well-known cloud-based cyberattacks that caused businesses to reevaluate their security measures.

Accenture Multiple Data Breaches: In 2017, Accenture let at least four of its AWS S3 storage buckets remain open, which resulted to a privacy breach that exposed sensitive information such as unrestricted authentication details, API data, digital certificates, decryption keys, user data, and meta information [34].

In August of 2021, Accenture was hit by another attack, carried out by the LockBit ransomware. During audits of the company's fourth and final quarter in 2021, the organization had the expertise this time to identify the intrusion.

Because of the data breach that occurred in 2021, Accenture anticipated that chain attacks were being carried out on client systems. These attacks involved caused problems in the critical systems, accidental exposure, and further malware infections. The perpetrators of the assault, the LockBit ransomware itself, claimed that they were able to steal 6 terabytes worth of data as a result of the attack and were demanding a ransom payment of $50 million.

Verizon Attacks: A misconfigured AWS S3 on the part of Nice Systems, a third-party partner of Verizon, led to the accidental disclosure of customer personally identifiable information in 2017 [35]. Because of an oversight made by Nice, the attack was able to take place. This fault allowed more customer call data to be captured. In the year 2020, Verizon discovered 29,207 security events, of which 5,200 were verified as breaches. The telecommunications behemoth fell victim to distributed denial-of-service assaults (DDoS). Each attack was powered by social engineering and client-side web app infections, which led to server-side system vulnerabilities.

The pandemic-induced remote productivity model is the key cause for the development of cyberattacks as well as the emergence of weaknesses in the system. The group classified these attacks as being the consequence of mistakes made by the "human factor", which is a byproduct of social engineering.

Kaseya Ransomware Attack: A significant cyberattack was launched against the unified remote monitoring and network perimeter security technology used by IT solutions company Kaseya in the month of July 2021 [36]. A ransomware operation along the supply chain had the objective of stealing administrative control of Kaseya services from managed service providers and the consumers they serve downstream. The incident rendered the company's SaaS servers inoperable and disrupted on-premise VSA systems that Kaseya clients in ten different countries were using. Kaseya's active response consisted of quickly notifying its clients. The firm released the Kaseya VSA detection tool, which enables corporate customers to examine the VSA services they utilize and monitor endpoints while looking for indicators of threats.

Cognyte: Cognyte, a large cybersecurity research company, made a mistake in May 2021 that resulted in its database being left unprotected and without authentication, paving the way for cyberattackers that exposed 5 billion user's data [37]. User credentials such as identities, email addresses, and passcodes, together with vulnerability data points inside their system, were compromised and disclosed. The data were accessible to the general public and were even included in web search indexes. Additionally, Cognyte's intelligence data, which include information about other data breaches of a similar kind, were made openly accessible to users. The data were secured by Cognyte over a period of four days.

Raychat: In 2021, A vulnerability in the database configuration allowed almost 267 million usernames, passwords, email addresses, metadata, and encrypted conversations to become public [39]. The entire company's data were deleted as a result of a coordinated bot attack. The business was keeping its customer information on a MongoDB database that was improperly configured. A NoSQL is a kind of storage application that is widely utilized by application firms to manage large amount of user information. However, if the NoSQL system is configured incorrectly, it can expose thousands of files to risk. In this particular instance, the malicious party managed to practically stroll right into the main doorway of Raychat and then execute a bot operation, which completely wrecked the database. The flaw was discovered by specialists while using open-source search tools that are utilized in the process of searching for gadgets that are linked to the internet. A large number of NoSQL databases, such as Mongo, are targets for BOT cyberattacks performed by malicious parties that scan the internet for open and unprotected DBS [databases] and delete their data, with just a ransom demand left behind.

Civicom: Amazon Simple Storage Service (Amazon S3) is a cloud-based storage solution that is scalable, has high transfer speeds, and is web-based. The service is created for use on Amazon Web Services with the aim of performing online backup and archiving of data and applications (AWS) [73]. The S3 bucket was left available to the public without a password or any other form of security authentication in October 2021 by Civicom, which meant that the data could be accessed by anyone who knew how to find corrupted databases. According to the findings of experts, Civicom disclosed 8 terabytes of documents, which included over 100,000 different files. This was caused by one of Civicom's Amazon S3 buckets having an inadequate configuration. In January 2022, after a period of three months, Civicom managed to secure the bucket.

FishPig: Malware was injected into Magento stores using a distribution network attack that is directed against the FishPig distribution server [74]. FishPig is a Magento extension supplier that has over 200,000 customers and specializes in Magento optimizations and interfaces between Magento and WordPress. This intrusion resulted in a threat actor introducing malicious PHP code into the Helper/License.php file. The injected code configures Rekoobe, which is another malicious program that hides itself as a background task inside systems that have been compromised. The malware introduced to License.php would download a Linux binary from license.fishpig.co.uk whenever the Fishpig control panel was accessed within the Magento system. The downloaded file, with the filename "lic.bin", pretends to be a licensing asset but is actually the Rekoobe remote access trojan. After being executed, the trojan deletes all harmful files from the infected machine but continues to operate in memory, imitating a system service, as it waits for commands from its command and control (C&C) server. The "lic.bin" file that was downloaded was actually a Rekoobe remote access trojan disguised as a licensing asset. The trojan removed all malicious files from the compromised computer but remained active in memory, masquerading as a system service while it awaits instructions from its command and control (C&C) server.

Microsoft: Microsoft reported on January 22, 2020 that one of its cloud databases had been accessed in December 2019, exposing 250 million emails, IP addresses, and support case information. The vulnerable customer service database was made up of a cluster of five Elasticsearch servers. This is a technology that is intended to facilitate search operations. Each of the five servers appeared to be a mirror image of each other since they all contained the same data. Microsoft claimed a faulty network server was the root cause of the data breach. Because of the prominent nature of the target, this incident was among one of the most alarming cyberattacks.

The following are a few observations that emerge from our investigation on root factors that contribute to the occurrence of most of the cyberattacks on cloud computing systems:

Misconfiguration: CSPs provide several levels of service to their customers, and these tiers are determined by the amount of control an organization requires over its cloud deployment. In order to achieve a higher level of cybersecurity, businesses need to design these installations according to the specifications of their operations. Unfortunately, a majority of firms do not have suitable cloud defense capabilities to guarantee the safety of these services, which leads to vulnerabilities in the deployment process. According to IBM [9], faulty server configurations are the root cause of 86% of all compromised data. Having knowledge of the particular deployment that the firm is using will make it easier to configure it according to the

company's unique security requirements using the security tools that
are supplied by CSPs.

Compromised user Accounts: The most common reason for user
accounts to be compromised is inadequate password policies. A
majority of users who use cloud services do not have adequate password security because they use weak passwords, reuse previous passwords, or do not frequently update their passwords. Therefore, users
should be strongly encouraged to change their passwords on a regular
basis, at least once every sixty to ninety days.

API Vulnerability: Users are able to communicate and collaborate
with their cloud-based computing platforms due to the availability of application programming interfaces (APIs) by cloud service
providers (CSPs). These application programming interfaces come
with a comprehensive documentation package, which enables users
to better comprehend and use the APIs. However, cybercriminals
are also able to get these documentations, and they may be used to
attack APIs in order to gain access to and remove sensitive data that
are kept in the cloud. Additionally, any flaws in the implementation
and setup of these APIs will make backdoors available for cybercriminals to get access to sensitive information. By strictly adhering to
the documentation, one can mitigate the risks of making any security
mistakes during the implementation and setup of APIs. Additionally,
organizations are required to carefully monitor the APIs' functionality in order to locate any potential vulnerabilities.

Insider Threats: A malicious user may bypass an organization's security policies and cause sensitive data to be compromised even if the
organization has implemented the most advanced cybersecurity environment possible. Because they may already have access to sensitive
data, malicious insiders' actions are sometimes difficult to identify. In
fact, the number of security breaches that have occurred as a consequence of insider threats has increased substantially over the course
of the last few years. Implementing strong access controls allows organizations to limit the amount of information that can be accessed by

persons working inside their own organization, which helps mitigate the risk posed by insider threats.

In today's cloud-driven interactions, users provide increasing amounts of their sensitive information to corporations, despite the fact that many of these corporations fail to appropriately safeguard these data. Because of the malicious techniques and technologies that are utilized to obtain private information, there are trust and confidentiality concerns. When a situation like this arises, authorities such as law enforcement, detectives, and system administrators turn to digital forensics for assistance in rearranging the sequence of events and locating traces of evidence that have been left behind.

4.   Cloud Forensic as a Form of Digital Forensics

Cloud forensics is a process that uses digital forensics techniques to investigate incidents that take place in the cloud. This technique is used to identify those responsible for the incidents. The use of digital forensic best practices inside a cloud context is what we mean when we talk about cloud forensics. According to NIST, Cloud computing forensic science uses scientific concepts, technology practices, and established procedures to recreate prior cloud computing events by identifying, acquiring, preserving, examining, interpreting, and reporting digital evidence [6].

After a crime has been committed, the forensic process may then begin. It discusses a variety of investigation methods [40], techniques, and tactics for certain crimes, in addition to the gathering of evidence. The current method of acquiring evidence has two major flaws: (a) the investigators need access to the suspect's stolen machine, and (b) traditional data storage, information exchange, and communication channels have been expanded by the newest emerging web services. Both of these flaws make the method less effective. It is possible for law enforcement agencies to request information from service providers; however, they are often not permitted to do so from other countries. As a consequence of this, the field of digital forensics is very new and is just in the beginning phases of its growth [41].

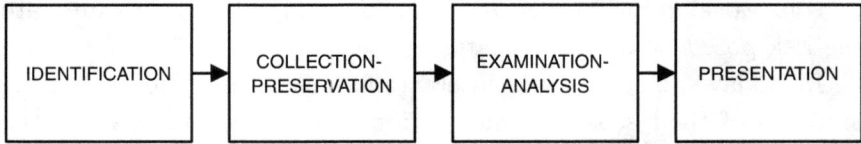

Figure 3. Four steps of digital forensics, proposed by authors.

Since 1999, a variety of strategies and frameworks have been developed, each consisting of a number of steps and phases, in order to successfully carry out digital forensic investigations. McKemmish [42] was one of the first researchers to describe forensic computing (actually creating the terminology digital forensics) as "the process of identifying, preserving, analyzing, and presenting digital evidence in a manner that is legally admissible". McKemmish was also one of the first researchers to coin the term "digital forensics". The four most important aspects (stages) of the forensic computing process are the identification, preservation, analysis, and presentation of digital evidence (Figure 3). During the identification phase of the investigation, the investigators are tasked with locating any and all potential sources of evidence. During the stage when preservation is taking place, the chain of custody needs to be kept intact at all times. The stage that needs expert testimony in a legal setting is the presenting stage, which contrasts with the stage that requires analysis, which comprises the extraction, processing, and interpretation of digital data.

5.   Dimensions of Cloud Forensics

NIST has put together a list of sixty-five cloud forensics difficulties. As a consequence of this, many industry professionals consider cloud computing to be an emerging platform for cyber-criminal activity [6]. The difficulties of how to perform a comprehensive digital investigation in cloud settings and how to plan to collect data ahead of time prior to the occurrence of an event are among these worries. Preparation of this kind would save money, effort, and time; thus, it is one of the primary concerns. On the basis of the evaluation of the

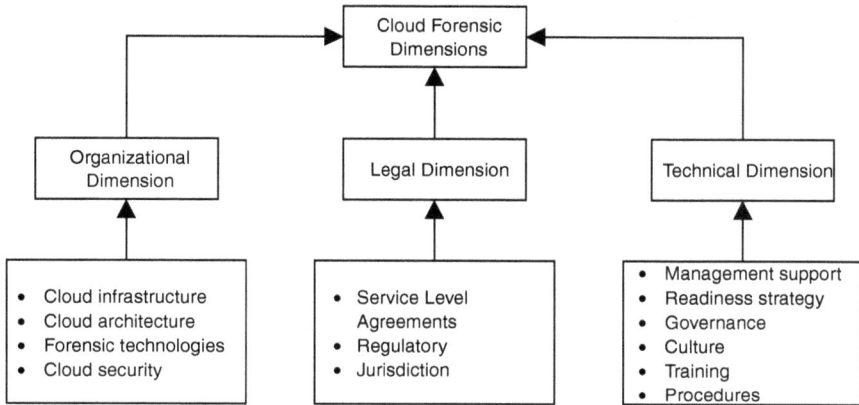

Figure 4. Dimension of cloud forensics.

relevant literature, the following framework is suggested: As shown in Figure 4, the structure comprises three distinct dimensions [71].

5.1. *Technical dimension*

Data gathering, live forensics, preventative measures, and virtualized settings are only some of the important tasks that have been completed. These tools are diverse from one another with regard to the deployment and service models that they use. Technical factors are support provided by management, forensic readiness techniques, governance, culture, training, and procedures.

5.2. *Legal dimension*

This is necessary to assure user privacy and that forensics procedures do not violate any laws or regulations in data center jurisdictions. Legal factors are service level agreements, regulatory, and jurisdiction.

5.3. *Organizational dimension*

Organizational Dimension: Both the cloud service provider and the cloud service consumers are members of the same organization.

When a cloud service provider makes its services available to users, the scope of the investigation expands. This study involves the participation of four distinct outside parties: cloud infrastructure, cloud architecture, forensic technologies, and cloud security.

6. Cloud Forensic Process Flow

The traditional steps of digital forensics, which can also be used in cloud forensics because of the different service and deployment models [72], are as follows:

6.1. *Identification*

Reporting on malicious activities is a form of identification. This occurs when a person or a CSP authority files a complaint on some incident. This process has two different types of identifications: identifying what happened and identifying what was found.

6.2. *Evidence collection*

According to forensic standards [44], the investigator gathers bits of evidence from cloud service models like SaaS, IaaS, and PaaS without putting the evidence's integrity at risk. SaaS service model looks at each user's data through log files like error log, access log, data volumes, application log, and transaction log. The IaaS service model looks at raw machine files, system logs, backups, storage logs, and other types of data. The PaaS service model looks at the data of application-specific logs though the API, operating system exceptions, malware software warnings, etc. All the evidence that has been gathered must be kept safe so that it can be used in future investigations. This can be done by sending court orders to cloud service providers. It is possible that keeping data safe will need a lot of storage space. The investigator talks about the rules for keeping data private and safe [45].

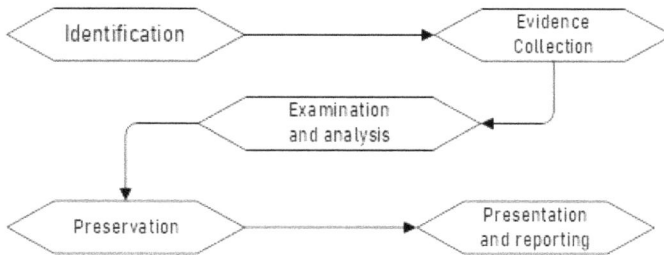

Figure 5. Cloud forensic process.

6.3. *Examination and analysis*

The analyst reviews the evidentiary information gathered by certain forensic tools in the previous stage. To get a logical conclusion, the criminal data are integrated, correlated, and assimilated. The analyst examines data from physical and logical files, regardless of where they are stored. It may be necessary to share the testimony with law police, a victim agency, or a person.

6.4. *Preservation*

All obtained evidence must be carefully stored and not tampered with for further investigation. Because the information is spread over many geographical locations, all log files must be saved.

6.5. *Presentation and reporting*

In the end, the investigator will put up a formal, well-organized report regarding the findings of the case so that they may be presented in a legal setting. The course of the cloud forensic method is shown in Figure 5.

7. Cloud Forensic Challenges

The cloud forensics faces a variety of challenges. We have discussed a number of difficulties associated with cloud computing forensics (Figure 6) in the following:

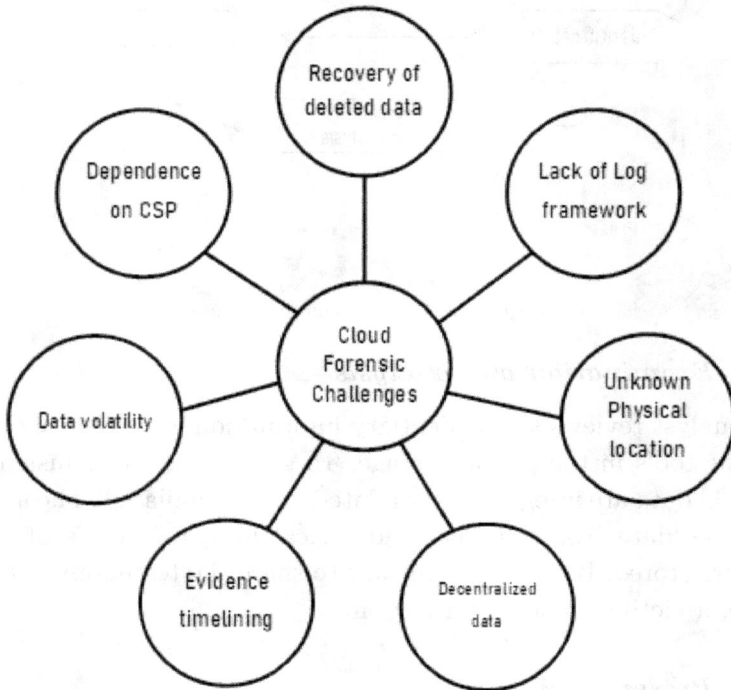

Figure 6.   Cloud forensic challenges.

- *Logs format unification*: Since the cloud computing environment consists of a huge number of servers, each of which has its own log format, this will make the investigation process more difficult. Because of the large number of servers that make up a cloud system and the many geographic locations at which these servers are located, each location operates under its own unique time zone.
- Intelligence processes for real-time investigation are often not possible in the cloud environment.
- Data integrity and evidence preservation.
- *Lack of terms and conditions in SLA*: Investigations have shown that the service level agreement (SLA), which should include the terms and conditions that govern the relationship between the user and the cloud service provider, is lacking terms and conditions.

These bullet points need to include significant language associated with the investigation of cloud forensics [47].

- Lack of forensics skills, particularly at the cloud computing level [48].
- There is a lack of coordination in cross-national data access and sharing, lack of international coordination, and regulatory mechanisms, especially when cloud forensics relies on gathering evidence from servers in several locations.
- To preserve evidence safely, strong encryption technologies are necessary [49].
- CSP reliance: Investigators rely on CSP to get logs. Other than CSPs, there is no proof linking a certain information file to a specific suspect.
- CSPs have agreements with other CSPs to utilize their services, which might result in data integrity and confidentiality being compromised in some situations.
- Forensics in the cloud are dependent on client–server communication in order to maintain their integrity and consistency. When data are transferred from the investigator's workstation to the cloud storage device, it is done so via a public network. Maintaining the integrity of the evidence is an essential part of the investigative process.
- It is possible that the evidence in cloud forensics will be destroyed. It becomes difficult to retrieve erased material and recreate it for use as evidence.

NIST states that "cloud computing" is an on-demand network access approach that allows users to quickly provide and release shared computing resources (e.g., networks, server storage, applications and services) with no administrative effort or service provider contact. Criminals, on the other hand, may take use of the properties of cloud computing in order to carry out illegal activities. Cybercriminals may exploit cloud computing as a topic, an object, or a tool [72]. Digital forensics in the cloud has a number of issues because of its diverse, virtualized, and distributed design [53]. In this way,

Table 3.   Challenges in Forensics phases in the cloud environment.

Steps of Digital Forensics	Challenges in Cloud Forensics
Identification	Decentralized data, unknown physical location, jurisdiction, data duplication, encryption, dependency chain, and dependency on CSP are all symptoms of this vulnerability.
Preservation	Evidence segregation, chain of custody, data volatility, distributed storage, and data integrity.
Collection	Absence of specialized commercial tools, dependency on a CSP, lack of control over data erasure policy, many tenants, and several jurisdictions all contribute to a lack of usability.
Examination and Analysis	Log framework, evidence time lines, encrypted data, and integration of evidence data.
Presentation and Reporting	Cloud model complexity and illiteracy, chain of custody, jurisdiction, and compliance.

new methods and tools may now be developed by researchers. The forensics steps in a cloud environment are also shown in Table 3.

8.   Cloud Forensic Models Addressing the Challenges

In this part of the chapter, we classify various cloud forensic research models according to the challenges outlined by NIST. It has been possible to build and use several cloud forensics methodologies that are customized to certain types of deployment and service needs. Customers in PaaS and SaaS service models, for example, have no control over the hardware and must depend on the CSP for logs, but customers in IaaS have the ability to generate an image of the instance and retrieve the logs themselves. For deployment techniques, public cloud customers lack accessibility and privacy compared to private cloud users. Virtualization brings the private cloud idea closer to the old local access networks that were used in the past. On-premises private cloud forensics are essentially identical to conventional forensics when it comes to conducting forensic investigations (internally). A forensic analysis of a private cloud hosted off site (external) is reliant on the CSPs and signed contracts. The most current research in cloud and digital forensics has been thoroughly examined, and the results are presented in this part in an in-depth study. This section includes ideas and theories from a variety of experts in digital and cloud forensics. To be clear, most of the works found are mainly focused on the investigation and resolution of cybercrime.

Case studies of widely used cloud storage systems such as DropBox, Amazon Cloud Drive, Amazon S3, and Google Docs were provided by a variety of authors to address the challenges of storage forensics. They discovered the possible evidence of artifacts in both the client-side as well as the server-side systems (cloud-native). Several articles [33,70,80,81] made use of the application programming interfaces (APIs) provided by the service provider in order to gather forensic artifacts with the intention of overcoming the limitations of client-side analysis.

Roussev *et al.* [28–30] raised doubts about the conventional method of client-side storage analysis and switched the emphasis to cloud-side analysis instead. They brought up crucial concerns such as artifact modifications, partial replication, and cloud-native artifacts in their presentations. They also placed a strong emphasis on an acquisition that was based on an API in order to solve the issues with client-side artifact analysis. On the other hand, Roussev *et al.* [29] created the kumodd tool as a proof-of-concept. This had three logical layers: the user interface, the dispatcher, and the drivers. In addition, the authors validated their findings by using the cloud storage services from Google Drive, 9 Microsoft OneDrive, 11 Dropbox, 10, and Box23. Later, Roussev *et al.* [30] pointed out a flaw in their previous work [29], that their kumodd tool lacked the ability to acquire cloud artifacts in their initial form due to the absence of API support. They developed a tool called kumodocs that is based on the Python programming language in order to investigate Google Docs artifacts. This tool is activated by a browser plugin called DraftBack24, which can replay the whole history of documents. The incompatibility of standard tools with StaaS (Storage as a service) applications was illustrated by Roussev *et al.* [28], who also presented 3 new forensic tools: kumodd, kumofs, and kumodocs.

Geographic locations can cause difficulty during the investigation procedure if it takes place in a cloud environment. Other Damshenas *et al.* [46] have suggested using a unified time-keeping system across all cloud entities as one of the potential solutions to the issue of time zones. This is one of the potential solutions that have been suggested. This has the benefit of establishing a logical order of events in terms of time.

Maintaining the integrity of the evidence is an essential part of the investigative process. There are many different solutions that have been proposed in order to improve the level of integrity of the cloud forensics investigation process. One such solution [50] suggests creating a digital signature for all of the collected evidence during the evidence gathering stage of the investigation, and then checking this signature on the other side prior to beginning the examination stage of the investigation. Hegarty [51] presented an additional solution to this problem, which offers a special framework for digital signature verification that enables forensic investigation of storage systems. The method was developed as an answer to this difficulty.

Zawoad *et al.* [52] proposed scheme based on SecLaas that gathers logs for permanent storage and limits the possibility of manipulation by using Proof of Past Log (PPL). It safeguards the privacy of cloud users by storing the logs generated by virtual machines and giving forensic investigators access to the data they need to do their investigations. In addition to this, SeclaaS stores evidence of prior logs, which defends the logs' authenticity against dishonest investigators or cloud service providers. They claimed that the anonymity of the information might be maintained if the investigators accessed the logs via RESTful APIs. In addition to this, they suggested a modification of the Bloom filter as well as the Bloom tree for the integrity verification process in order to obtain increased performance in terms of both time and space.

Rane and Dixit [57] used the irreversible nature of blockchain to maintain the secrecy and authenticity of cloud logs, and they offered secure logging-as-a-service for the cloud environment. This was accomplished by using a distributed ledger. Steps such as extracting logs from a virtualized environment, creating encrypted log entries for each log using public key encryption, and storing encrypted log entries on the blockchain are all included in the system. Similarly, Wang *et al.* [58] suggested a public key verification approach based on blockchain that would use a third-party inspector to certify the authenticity of logs stored in cloud storage. The authors made use of homomorphic (calculating the hash of a composite block by adding the hashes of the separate blocks together) and one-way hash functions in order to produce labels for log entries, and they chose to store their data in a Merkel tree structure.

To address the logs format unification challenge, a log format unification on cloud services with the use of the Cloud Auditing Data Federation (CADF) standard developed by the Distributed Management Task Force (DMTF) [54]. OpenStack makes use of CADF event logging, and the authors have updated the Apache CloudStack platform to make it forensically sound. They also analyzed the preexisting CloudStack platform in addition to the suggested CADF event model that was implemented.

Ahsan *et al.* [55] developed cloud log assuring soundness and secrecy scheme (CLASS) for cloud forensic where the public key of user was used to encrypt the logs. They used Rabin's fingerprint and bloom filter to avoid unauthorized modification of the log. This approach reduced the time needed for verification.

An open-source framework was designed by Deshpande and Rao [56] using open-source tools such as apache flume, apache kafka, ELK stack, and apache spark for real-time and historical log analytics which can assist the system admins to monitor critical incidents while analyzing in real time.

Because of the virtualization, distribution, and dynamic nature of cloud systems, developing a cloud architecture that can enable forensics is a huge and difficult challenge. This challenge involves a large number of delicate legal, organizational, and technological challenges. Hemdan and Manjaiah [59] offered an effective Cloud Forensics Investigation Model (CFIM) that can investigate crimes committed in the cloud in a manner that is both forensically sound and timely.

The system is compatible with the idea of Forensic as a Service (FaaS), which offers a multitude of advantages to the process of doing digital forensics by using a forensic server on the cloud side.

To ensure the trust and decrease the dependency on CSPs, Alex and Kishore [60] has offered a model to improve cloud forensics. In this model, both the forensic monitoring plane (FMP) and the forensic server have been developed. Forensics tools such as the forensic toolkit (FTK) analyzer and E-Detection running at the top of the FMP will supervise both inbound and outbound links in the cloud system. The data that are monitored are encrypted bit by bit stream and stored on a separate forensics server that is positioned at the crime site. The forensic tool monitors how cloud service models behave. Therefore, when an event takes place in a particular cloud, all of the actions that take place in that cloud, including network traffic, are forensically imaged. The image is encrypted again and stored in the forensic server which enables a reduction in the trust amount placed in the CSP. In the event that there was any kind of malicious behavior, the investigator may immediately connect to the forensic server using their user credentials and gather forensic evidence within a reasonable amount of time after the incident. In the meanwhile, if the investigator has any reason to suspect anything, they may make a data request to CSP and compare it to the information they have gotten from the forensic server.

A log assuring confidentiality mechanism suitable for use in the edge-cloud environment was suggested by Park and Huh [61]. The investigators will attempt to retrieve the log files using this method. In the beginning, each and every log will be saved on one of the edge nodes. After the segmentation has been saved to the centralized cloud node and the distributed storage system, the users will log in. During the initial phase of the attack, the attacker will attempt to delete or steal data from the edge node rather than the central store or the distributed storage system. In the second step of the attack, the attacker will attempt to steal data by attacking the cloud node. Meanwhile, CSP will communicate with the investigator so that they can investigate and store the data. The investigator is able to retrieve the data from the central storage or distributed storage system, but the edge node, which was either taken or destroyed by the attacker, is beyond their ability to recover. In addition, the data may be eventually retrieved from the distributed storage clusters with the assistance of the MIC network and index files.

Raju *et al.* [62,63] sought to solve the problems that arise while attempting to reconstruct events in a cloud context due to multi-tenancy and privacy violations, various heterogeneous data sources, and a huge number of events. They suggested frameworks and techniques for successful cloud event restoration based on the log aggregation algorithm Leader–Follower (LF) [62] and classical digital event restoration [63].

A forensic pattern-built technique was developed in 2018 by Juan-Carlos Bennett and Mamadou H. Diallo [66]. This approach is a semiformal architecture that is based on an object-oriented approach in the context of patterns. They have been using the NIST Forensics Framework throughout the whole process of collecting, reviewing, and analyzing the evidence. They developed the Cloud Evidence Collector in addition to the Cloud Evidence Analyzer so that they could collect more and better evidence of the network in the shortest period of time feasible.

Although it may not be commonly explored in earlier researches, log correlation is an important factor to take into account when doing analysis of cloud logs. For instance, Marty [43] concentrated only on the logging of certain programs and did not take into account the connection between the various logs. The prototype implementation that is described in reference [64] was created for Windows event logs, and the safety of the centralized log server was not taken into consideration throughout its development. Kernel change [69] for forensically enabled cloud environment may be expensive because of the size of the cloud's hardware infrastructure, and it may not be practical for cloud service providers (CSPs) since it would necessitate the shutdown and restart of existing cloud services. The method that Darren Quick and Kim-Kwang Raymond Choo [70] used was one in which the authors did not show any functioning prototype of their proposed notion and instead made the assumption that perhaps the CSP can be trusted completely.

Park [75] proposed a forensics framework that is a blockchain-based data storage and integrity management mechanism. Since blockchain maintains the integrity of data, this framework should be used to store forensic evidence. As a result of the fact that all of the blocks are linked to one another, the integrity of the data can be easily checked by using the hash value of the block that comes before it.

Protecting the logs is yet another crucial component of the process [76]. Cloud maintains a record of each and every action taken inside its environment. On the basis of the log data, a forensics expert conducts investigation into the incident. A forensic investigator's primary concern should be protecting the users' privacy wherever possible. This is necessitated by the fact that cloud computing is utilized continuously by plenty of users, all of whom share the same storage and processing network.

CSPs have to be available to the third-party investigators for inspection of logs. In order to solve this problem, there is a log block tag system [77] that is based on the Merkle Hash Tree.

Pourvahab [78] designed a blockchain system that operates in the cloud with IaaS concept. The forensic design of the blockchain, in which peers are scattered among different nodes, is used in this method for the purpose of gathering and storing evidence. It was intended that secure ring verification-based authentication would be used for the purpose of safeguarding a device from being used by unauthorized parties.

Maintaining the integrity of the evidence is an essential part of the investigative process. There are many different solutions that have been proposed in order to improve the level of integrity of the cloud forensics investigation process. One such solution [50] suggests creating a digital signature for all of the collected evidence during the evidence gathering stage of the investigation, and then checking this signature on the other side prior to beginning the examination stage of the investigation. Hegarty [51] presented an additional solution to the concern related to integrity, which offers a special framework for digital signature verification that enables forensics investigation of storage systems.

9. Cloud Specific Forensic Tools

The development of digital forensic tools is still in the early stages due to the ever-changing nature of new technologies and the ever-increasing number of devices that utilize these technologies. The National Institute of Standards and Technology (NIST) produced a database of numerous digital forensics tools and their capabilities [67]. However, the category for cloud services only contains a set of six tools altogether. In this area, we classify forensics tools into two categories: cloud-specific solutions and classic digital forensics tools [68]. There is a list of these instruments in Table 4.

10. Opportunities of Cloud Forensics

Cost Efficiency: Compared to cloud computing, which is more cost-effective and less expensive when deployed at a bigger size, cloud forensics is also less expensive when executed on a larger scale.

Abundance of Data: Due to the fact that the data are replicated and stored across multiple data centers and servers, it is possible that the data have not been completely deleted. These data may be available for forensics, and even if deleted, it may still be possible to recover it. This is due to the fact that the data are replicated and stored across multiple data centers and servers.

Regulations and guidelines: Because cloud computing is still in its development era, it is the ideal moment for cloud forensics to establish the groundwork for their policies and standards.

Forensics as a Service: Forensics as a Service, also known as FaaS, is something that can be created, and once it is, it may be of great

Table 4. Conventional digital forensic tools.

Tools	Functions	Service Model/OS
	Digital Forensic Tools	
DFF [25]	Instrument used in forensic investigation to detect, acquire, and store evidence while maintaining a chain of custody.	IaaS
EnCase Forensic [31]	A forensic solution in the form of a set of software that may be used to gather, preserve, analyze, and report on evidences in a manner that is acceptable to the court.	IaaS
Access Data Forensic Toolkit (FTK) [25,31]	It is a collection of forensic tools, such as FTK imager for disc imaging, that may be used to analyze emails, carve data, and do other tasks.	IaaS
Wireshark [31]	Network protocol analyzer	IaaS
Wild packets Omnipeek40 [21]	Enterprise-level software for the examination of network packets and protocols.	All
eNetwork Miner [10]	Open-source Network Forensics Analysis Tool (NFAT).	All
X-Ways Forensic [11]	Integrated forensic environment with a range of capabilities such as access to file system structures with deleted partitions, disc imaging and cloning, and file and directory catalog.	SaaS
	Cloud Specific Tools	
FROST [38]	A forensic toolset built for the OpenStack cloud platform that can collect forensic evidence independently of a cloud service provider (CSP).	IaaS
PALADIN [12]	This digital forensic program offers more than one hundred helpful tools that may be used to analyze any potentially harmful content.	Windows and Linux
Kumodd [31,59]	A forensic tool for cloud storage that can acquire cloud drives and take snapshots of cloud-native artifacts in formats such as PDF.	SaaS
e-Fencer [13]	It provides investigators with a straightforward interface for searching for data on whatever device they want.	Windows, Mac OS X, and Linux
Kumodocs [59]	Analysis software for Google Docs relying on Draft-Back, a browser plugin that replays the whole history of documents sitting in the Document folder. Draft-Back was developed by Google.	SaaS

Table 4. (*Continued*)

Tools	Functions	Service Model/OS
Kumofs [59]	Tool used in forensic investigations for the gathering and examination of file metadata that are stored in the cloud.	SaaS
VNsnap [8]	Cloud-based snapshot tool designed for use with virtual network architectures.	IaaS
EnCase [15]	The ability to recover credentials from the hard disc is facilitated for the investigators by this. It enables investigators to conduct in-depth analyses of case files in order to collect evidence such as papers and images.	Windows
Crowdstrike [16]	• Backing up virtual, physical and cloud-based data centers. • Managing system errors. • Automatically detecting malware.	Windows and Mac
Xplico [17]	• Data output into MySQL or SQLite database. • Reserves DNS search from DNS packages that contain input files. • Gives features like Port Protocol Identification (PPI) to assist digital forensics. • Open source and supports IPv4 as well as IPv6.	Linux
SANS SIFT [18]	• Installable using SIFT CLI. • Improves memory. • Latest forensic methods and tools	Windows, Linux, Mac OS X,
Cloud Data Imager [33]	Innovative solution for the remote acquisition of cloud storage that primarily consists of two features: directory browsing and the creation of a logical duplicate of the folder tree that has been picked.	SaaS
LINEA [19]	A forensic instrument for the gathering of live network evidence from various web services.	SaaS
ForenVisor [68]	A piece of software in the form of a dynamic hypervisor that can do live forensic analysis.	IaaS

support in solving crimes linked to the cloud, as well as other types of cybercrime investigations.

11. Future Work

Cloud forensics investigators must collect all evidence from distributed cloud architecture. Future research can focus on building a reliable and effective framework for investigators to avoid errors in evidence collection and integration. Methods of machine learning, such as log correlation and meta-data analysis, are becoming an increasingly important focal point in cloud computing. Some stages of cloud forensics, such as federated learning, may benefit from the use of automation [65]. Cloud forensics datasets that include cloud meta-data attributes are not publicly available. Contributing to the cloud forensics dataset in the future will include acquiring evidence data, preserving user privacy, and adhering to new laws, all of which require current legal knowledge as well. The researchers are moving in the direction of standardization as the ultimate objective of cloud forensics. The provision of an all-encompassing cloud forensics solution is one way to achieve standardization.

12. Conclusion

Nowadays, cloud computing is one of the most important issues in information technology. Everything is on the cloud and people can easily access it from anywhere through the internet. Therefore, all the giant organizations are using these services or are planning to move toward it. Any new development in the cloud will almost certainly uncover some previously unknown dangers. We believe this is mostly due to increasing competence in the field as well as growing acceptance of cloud computing within the information and communications technology sector. The field of cloud forensics is fraught with difficulties at every stage of the investigation process. This comprehensive literature review demonstrates that cloud forensics is moving in the direction of evidence provenance. In addition to that, the study demonstrates how the various aspects of cloud forensics may have an effect on the forensics process. In this study, we provided a unique cloud forensic taxonomy that addresses a number of challenges related to evidence collection, analysis, trust, and legal complications. In addition to this, this chapter makes it abundantly clear that future research might concentrate on developing a dependable and efficient framework for investigators to use in order to prevent mistakes in the process of evidence collecting and integration. An effective contribution to cloud forensics may be made through the intelligent gathering of prospective evidence. Despite the fact that trust problems cannot be eliminated entirely, the use of a system that is based on provenance may give the necessary level of confidence for the inquiry.

References

[1] Ruan, K., Carthy, J., Kechadi, T., and Crosbie, M. Cloud forensics. In *An Overview IFIP Conference on Digital Forensics*, pp. 35–46, 2011. https://doi.org/10.1007/978-3-642-24212-03.

[2] Buyya, R., Yeo, C.S., Venugopal, S., *et al.* Cloud computing and emerging IT platforms: Vision, hype, and reality for delivering computing as the 5th utility. *Future Generation Computer Systems*, 25:599–616, 2009. https://doi.org/10.1016/j.future.2008.12.001.

[3] Raj Vardhman. Cloud Computing Statistics and Facts – 2021. Published on: 7 June 2021.

[4] De Marco, L., Kechadi, M.-T., and Ferrucci, F. Cloud forensic readiness: Foundations. In *International Conference on Digital Forensics and Cyber Crime*, pp. 237–244, 2013.

[5] Market Research Media, 2016. http://www.marketresearchmedia.com/?p=839. Accessed 16 March 2022.

[6] NIST. NIST Cloud Computing Forensic Science Challenges, 2020.

[7] Cloud Security Alliance (CSA): Mapping the Forensic Standard ISO/IEC 27037 to Cloud Computing, 2021. 6.

[8] Ruan, K., James, J., Carthy, J., and Kechadi, T. Key terms for service level agreements to support cloud forensics. In *IFIP International Conference on Digital Forensics*, pp. 201–212, 2019.

[9] Configuration mistakes blamed for bulk of stolen records last year: IBM. Available Online: https://www.itworldcanada.com/article/configuration-mistakes-blamed-for-bulk-of-stolen-records-last-year-ibm/427178. Accessed on 19 September 2022.

[10] Spiekermann, D., Eggendorfer, T., Keller, J. Using network data to improve digital investigation in cloud computing environments. In *2015 International Conference on High Performance Computing & Simulation (HPCS)*, 20 July 2015, pp. 98–105. IEEE.

[11] Bhagat, S.P., Meshram, B.B. Digital forensic tools for cloud computing environment. In *ICT with Intelligent Applications*, pp. 49–57. Springer, Singapore, 2022.

[12] The World's Most Popular Linux Forensic Suite. https://sumuri.com/software/paladin/. Accessed on 7 September 2022.

[13] Don't let your company data walk out the door! http://www.e-fense.com/products.php. Accessed on 20 September 2022.

[14] da Silva, C.M.R., da Silva, J.L.C., Rodrigues, R.B., Campos, G.M.M., do Nascimento, L.M., Garcia, V.C. Security threats in cloud computing models: Domains and proposals. In *2013 IEEE Sixth International Conference on Cloud Computing*, June 2013, pp. 383–389.

[15] OpenText EnCase Forensic. https://security.opentext.com/encase-forensic. Accessed on 7 September 2022.

[16] https://www.crowdstrike.com/endpoint-security-products/falcon-endpoint-protection-pro/. Accessed on 20 September 2022.

[17] https://www.xplico.org/. Accessed on 20 September 2022.

[18] https://digital-forensics.sans.org/community/downloads/. Accessed on 20 September 2022.

[19] Kaur, M., and Singh, H. A review of cloud computing security issues. *International Journal of Advanced Engineering Technology*, 8(3):397–403, 2016.

[20] Federici, C. Cloud data imager: A unified answer to remote acquisition of cloud storage areas. *Digital Investigation*, 11(1):30–42, 1 March 2014.

[21] Khalil, I., Khreishah, A., and Azeem, M. Cloud computing security: A survey. *Computers*, 3(1):1–35, 2019.

[22] Quick, D., and Choo, K.K. Forensic collection of cloud storage data: Does the act of collection result in changes to the data or its metadata? *Digital Investigation*, 10(3):266–277, 1 October 2013.

[23] Roussev, V., Barreto, A., and Ahmed, I. Forensic acquisition of cloud drives. arXiv preprint arXiv:1603.06542. 26 January 2016.

[24] Metheny, M. Federal cloud computing. Syngress, 2017.

[25] Walker, K. Cloud security alliance (CSA). The treacherous 12: cloud computing top threats in 2016, 29 February 2016. https://cloudsecurityalliance.org/media/news/cloud-security-alliance-releases-the-treacherous-twelve-cloud-computing-top-threats-in-2016/.

[26] Grobauer, B., Walloschek, T., and Stöcker, E. Understanding cloud-computing vulnerabilities. *IEEE Security and Privacy*, 2020.

[27] Roussev, V., and McCulley, S. Forensic analysis of cloud-native artifacts. *Digital Investigation*, 16:S104–S113. 29 March 2016.

[28] Roussev, V., Ahmed, I., Barreto, A., McCulley, S., Shanmughan, V. Cloud forensics–Tool development studies & future outlook. *Digital Investigation*, 18:79–95. 1 September 2016.

[29] Roussev, V., Barreto, A., and Ahmed I. Forensic acquisition of cloud drives. arXiv preprint arXiv:1603.06542. 26 January 2016.

[30] Roussev, V., and McCulley, S. Forensic analysis of cloud-native artifacts. *Digital Investigation*, 16:S104–S113. 29 March 2016.

[31] Yan, L., Rong, C., and Zhao, G. Strengthen cloud computing security with federal identity management using hierarchical identity-based cryptography. *Cloud Computing*, pp. 167–177, 2019.

[32] Brunette, G., and Mogull, R. Security Guidance for Critical Areas of Focus in Cloud Computing V2. 1. CSA (Cloud Security Alliance), USA.

[33] Khorshed, M.T., Ali, A.S., and Wasimi, S.A. Trust issues that create threats for cyber-attacks in cloud computing. In *2011 IEEE 17th International Conference on Parallel and Distributed Systems*, 7 December 2011, pp. 900–905. IEEE.

[34] System Shock: How a Cloud Leak Exposed Accenture's Business | UpGuard. Available online: https://www.upguard.com/breaches/cloud-leak-accenture.

[35] 2022 Data Breach Investigations Report | Verizon. Available online: https://www.verizon.com/business/resources/reports.

[36] Updated Kaseya ransomware attack FAQ: What we know now | ZDNET. Available online: https://www.zdnet.com/article/updated-kaseya-ransomware-attack-faq-what-we-know-now/.

[37] Top 5 Cloud Security Data Breaches in Recent Years (makeuseof.com). Available online: https://www.makeuseof.com/top-recent-cloud-security-breaches/.

[38] Mathisen, E. Security challenges and solutions in cloud computing. In *5th IEEE International Conference on Digital Ecosystems and Technologies (IEEE DEST 2011)*, 2021.

[39] Iranian Chat App Gets Its Data Wiped Out in a Cyberattack. Available online: gizmodo.com/iranian-chat-app-gets-its-data-wiped-out-in-a-cyberatta-1846181651.

[40] Al Sadi, Ghania. Cloud computing architecture and forensic investigation challenges. *International Journal of Computer Applications*, 124(7), 2015.

[41] O'shaughnessy, Stephen, and Anthony Keane. Impact of cloud computing on digital forensic investigations. *IFIP International Conference on Digital Forensics*, Springer, Berlin, Heidelberg, 2013.

[42] McKemmishR. What is forensic computing? Australian Institute of Criminology. Canberra, 1999, p. 118.

[43] Marty, R. Cloud application logging for forensics. In *Proceedings of the 2011 ACM Symposium on Applied Computing*, 21 March 2011, pp. 178–184.

[44] Wang, Ping, *et al.* Clustering-based emotion recognition micro-service cloud framework for mobile computing. *IEEE Access*, 8:49695–49704, 2020.

[45] Djemame, Karim, *et al.* PaaS-IaaS inter-layer adaptation in an energy-aware cloud environment. *IEEE Transactions on Sustainable Computing*, 2(2):127–139, 2017.

[46] Damshenas, M., Dehghantanha, A., Mahmoud, R., and bin Shamsuddin, S. Forensics investigation challenges in cloud computing environments. In *2012 International Conference on Cyber Security, Cyber Warfare and Digital Forensic (CyberSec)*, 26 June 2012, pp. 190–194. IEEE.

[47] Ruan, K., Carthy, J., Kechadi, T., and Crosbie, M. Cloud forensics. In *IFIP International Conference on Digital Forensics*, 31 January 2011, pp. 35–46. Springer, Berlin, Heidelberg.

[48] Khanafseh, M., Qatawneh, M., and Almobaideen, W. A survey of various frameworks and solutions in all branches of digital forensics with a focus on cloud forensics. *International Journal of Advanced Computer Science and Applications*, 10(8):610–629, 2019.

[49] Agarkhed, J., and Ashalatha, R. An efficient auditing scheme for data storage security in cloud. In *2017 International Conference on Circuit, Power and Computing Technologies (ICCPCT)*, 20 April 2017, pp. 1–5. IEEE.

[50] Zawoad, S., and Hasan, R. Cloud forensics: A meta-study of challenges, approaches, and open problems. arXiv preprint arXiv:1302.6312. 26 February 2013.

[51] Hegarty, R., Merabti, M., Shi, Q., and Askwith, B. Forensic analysis of distributed data in a service-oriented computing platform. In *Proceedings of the 10th Annual Postgraduate Symposium on the Convergence of Telecommunications, Networking & Broadcasting, PG Net*, June 2009.

[52] Zawoad, S., Dutta, A.K., and Hasan, R. Towards building forensics enabled cloud through secure logging-as-a-service. *IEEE Transactions on Dependable and Secure Computing*, 13(2):148–162. 25 September 2015.

[53] Ahmed Ali, S., Memon, S., and Sahito, F. Challenges and solutions in cloud forensics. In *ACM International Conference Proceeding Series*, pp. 6–10, 2018. doi: 10.1145/3264560.3264565.

[54] Dalezios, N., Shiaeles, S., Kolokotronis, N., and Ghita, B. Digital forensics cloud log unification: Implementing CADF in Apache CloudStack. *Journal of Information Security and Applications*, 54:102555. 1 October 2020.

[55] Ahsan, M.M., Wahab, A.W., Idris, M.Y., Khan, S., Bachura, E., and Choo, K.K. Class: Cloud log assuring soundness and secrecy scheme for cloud forensics. *IEEE Transactions on Sustainable Computing*, 6(2):184–196. 7 May 2018.

[56] Deshpande, K., and Rao, M. An open-source framework unifying stream and batch processing. In *Inventive Computation and Information Technologies*, pp. 607–630. Springer, Singapore, 2022.

[57] Rane, S., and Dixit, A. BlockSLaaS: Blockchain assisted secure logging-as-a-service for cloud forensics. In *International Conference on Security & Privacy*, 9 January 2019, pp. 77–88. Springer, Singapore.

[58] Wang, J., Peng, F., Tian, H., Chen, W., and Lu, J. Public auditing of log integrity for cloud storage systems via blockchain. In *International Conference on Security and Privacy in New Computing Environments*, 13 April 2019, pp. 378–387. Springer, Cham.

[59] Hemdan, E.E., and Manjaiah, D.H. An efficient digital forensic model for cybercrimes investigation in cloud computing. *Multimedia Tools and Applications*, 80(9):14255–14282. April 2021.

[60] Alex, M.E., and Kishore, R. Forensics framework for cloud computing. *Computers & Electrical Engineering*. 60:193–205. 1 May 2017.

[61] Park, J., and Huh, E.N. eCLASS: Edge-cloud-log assuring-secrecy scheme for digital forensics. *Symmetry*, 11(10):1192. 22 September 2019.

[62] Raju, B.K., Gosala, N.B., and Geethakumari, G. Closer: Applying aggregation for effective event reconstruction of cloud service logs. In *Proceedings of the 11th International Conference on Ubiquitous Information Management and Communication*, 5 January 2017, pp. 1–8.

[63] Kumar Raju, B.K., and Geethakumari, G. Timeline-based cloud event reconstruction framework for virtual machine artifacts. In *Progress in Intelligent Computing Techniques: Theory, Practice, and Applications*, pp. 31–42. Springer, Singapore, 2018.

[64] Trenwith, P.M., and Venter, H.S. Digital forensic readiness in the cloud. In *2013 Information Security for South Africa*, 14 August 2013, pp. 1–5. IEEE.

[65] Valjarevic, A., and Venter, H.S. Harmonised digital forensic investigation process model. In *Information Security for South Africa (ISSA)*. IEEE, Johannesburg, Gauteng, 2012.

[66] Bennett, J.C., and Diallo, M.H. A forensic pattern-based approach for investigations in cloud system environments. In *2018 2nd Cyber Security in Networking Conference (CSNet)*, 24 October 2018, pp. 1–8. IEEE.

[67] National Institute of Standards and Technology, 2018. Computer Forensics Tool Catalog. Retrieved from https://toolcatalog.nist.gov.

[68] Manral, B., Somani, G., Choo, K.K., Conti, M., and Gaur, M.S. A systematic survey on cloud forensics challenges, solutions, and future directions. *ACM Computing Surveys (CSUR)*, 52(6):1–38. 14 November 2019.

[69] Trenwith, P.M., and Venter, H.S. Digital forensic readiness in the cloud. In *2013 Information Security for South Africa*, 14 August 2013, pp. 1–5. IEEE.

[70] Darren Quick and Kim-Kwang Raymond Choo. Forensic collection of cloud storage data: Does the act of collection result in changes to the data or its metadata? *Digital Investigation*, 10(3): 266–277, 2013.

[71] Alenezi, A., Hussein, R.K., Walters, R.J., and Wills, G.B. A framework for cloud forensic readiness in organizations. In *2017 5th IEEE International Conference on Mobile Cloud Computing, Services, and Engineering (MobileCloud)*, 6 April 2017, pp. 199–204. IEEE.

[72] Datta, S., Majumder, K., De, D. Review on cloud forensics: An open discussion on challenges and capabilities. *International Journal of Computer Applications*, 145(1):1–8, 2016.

[73] Market Research and Conferencing Service Civicom Leaked 8TB of Data. Available Online: https://cyberintelmag.com/cloud-security/market-research-and-conferencing-service-civicom-leaked-8tb-of-data/. Accessed 17 September 2022.

[74] Malware Spreads Through FishPig Distribution Server. Available Online: cyberintelmag.com/malware-viruses/malware-spreads-through-fishpig-distribution-server-to-infect-magento-powered-stores/. Accessed 17 September 2022.

[75] Alharby, M., and Van Moorsel, A. Blockchain-based smart contracts: A systematic mapping study. arXiv preprint arXiv:1710.06372. 17 October 2017.

[76] Wang, J., Peng, F., Tian, H., Chen, W., and Lu, J. Public auditing of log integrity for cloud storage systems via blockchain. In *International Conference on Security and Privacy in New Computing Environments*, 13 April 2019, pp. 378–387. Springer, Cham.

[77] Kaleem, H., and Ahmed, I. Cloud forensics: Challenges and solutions (blockchain based solutions). *Innovative Computing Review*, 1(2):1–26. 26 December 2021.

[78] Pourvahab, M., and Ekbatanifard, G. Digital forensics architecture for evidence collection and provenance preservation in IaaS cloud environment using SDN and blockchain technology. IEEE Access, 7:153349–153364. 11 October 2019.

https://doi.org/10.1142/9789811273551_0005

Chapter 5

Smart City Vulnerabilities: An Overview

Amardeep Das[*,†,§] **and Pradeepkumar Bhale**[‡,¶]

Utkal University, Vanivihar, Bhubaneswar, Odisha, India
†*C.V. Raman Global University, Bhubaneswar,*
Odisha, India
‡*Indian Institute of Technology Guwahati,*
Guwahati, Assam, India

§*amardeepcvrp@gmail.com*
¶*pradeepkumar@iitg.ac.in*

Abstract: With recent improvements in information and communication technology, a "smart city" has been created to make the most of the resources in cities in a dynamic way. Smart cities can provide new applications and services to enhance their residents' lives in energy usage, transportation, healthcare, and education. However, they create new vulnerabilities and threats, including making city infrastructure and services insecure, brittle, and open to extended criminal activity. Commercial and government interests have mainly disregarded or underestimated this contradiction or used a technically mediated mitigating technique. This chapter provides a detailed review of various vulnerability challenges in smart cities and a foundation for classifying current and future advancements in this field. In addition, it describes the security requirements for designing a security solution for a smart city, identifies the existing security solutions, and discusses the open research issues and challenges of smart city security.

Keywords: Smart City, Vulnerabilities, Security, Privacy, Internet of Things (IoT).

1. Introduction

In recent years, cities have adopted more technology and become more intelligent. With the assistance of emerging technologies and fast, easy communication, cities can make much better use of their resources, save money, and give their citizens better services [1]. The world's governments have realized that although they regularly face problems in achieving their goals due to restricted budgets, insufficient resources, and antiquated systems, emerging technology can convert those difficulties into opportunities [2]. According to [3], a smart city utilizes systems to automate and change governmental tasks and enhance the lives of its residents. With digital technology, a "Smart City" may improve the quality of municipal services, cut costs and waste, and enhance public engagement and participation [4]. Government services, transportation, electricity, water, health, and waste management are just a few of the areas that have flourished because of the introduction of smart city technology. The Internet of Things (IoT) sensors come in various forms, and a smart city includes all of them. Some examples of the applications of these sensors are smart parking, traffic control, lane optimization, smart lighting, and structured health awareness [5]. The IoT is an operational technology employed for those smart city parts. The cloud is a useful platform to store and analyze centralized smart city data [3]. As shown in Figure 1, smart city architecture may comprise the following components:

I. **Smart government:** Smart government is the deployment of business procedures that use information and communication technology (ICT) to provide information continuity between government and the delivery of high-quality services. The smart government uses real-time information to cut down on crime by making people more aware of their surroundings, responding quickly and effectively to accidents, and improving city services.

II. **Smart healthcare:** Smart healthcare is a type of medical service that connects people, places, and organizations by utilizing

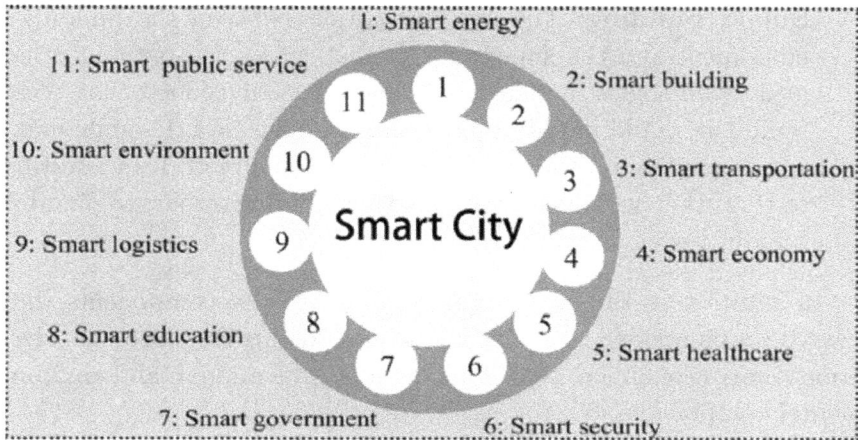

Figure 1. Architecture of smart city [3].

the IoT. The fundamental elements of smart healthcare include doctors, patients, hospitals, and medical research facilities.

III. **Smart energy:** The infrastructure of the conventional electricity system cannot keep up with the expanding demands of communities. As a result of instantaneous monitoring made possible by the smart grid, clients and the power grid receive optimal power flows. Additionally, by incorporating renewable energy sources into the grid, it makes it possible to produce electricity that is ecologically benign (for both the power company and consumers).

IV. **Smart transportation:** One of the ultimate aims of traffic management systems is to enhance the efficiency of networks while simultaneously increasing the safety of both humans and vehicles. There are five categories that may be used to categorize intelligent transportation systems: proactive road safety, location-based services, cooperative traffic efficiency, traffic psychology, and worldwide Internet services. The most significant benefits of utilizing intelligent transportation systems are a decrease in the amount of time spent in traffic and an increase in the overall level of safety.

V. **Smart building:** The communication between the building's
equipment and the smart grid is enabled by sensors and other
grid technologies in smart buildings. It is envisioned that these
buildings would modify their energy profiles in a dynamic manner depending on the capabilities of the smart grid. In addition
to that, they will make it possible for building owners to monitor
building systems remotely.

In addition to the architecture and smart city components that
have been mentioned so far, it is also possible to take into consideration smart economics, security, logistical, educational, and environmental components [6].

1.1.   *Operational architecture of smart cities*

Typically, the operational architecture of smart cities is one of a kind,
and it is composed of four separate levels, each of which has certain
roles and components. These layers are as follows:

  I. **Sensing layer:** This is the initial layer of the smart city architecture, and it is made up of a variety of various components
and instruments to collect data from the world around it. Some
examples of these components are cameras, actuators, and sensors of varying types. In order to gather data and send it to
the data collection layer, the sensor nodes need to be placed in
various parts of the smart city. The operational architecture of
smart cities is depicted in Figure 2, with the key elements of
each layer.

 II. **Data collection layer:** This layer involves the collection of data
from the various resources of smart cities, including homes, traffic, and citizens. It is necessary to transmit the data using trusted
wired or wireless communication to the regional databases in
charge of storing the collected data. However, this layer's present
databases have a substantial storage cost, making it impractical
to save such a large amount of data. While most architectural
solutions include storing data on the cloud, we still need efficient parsing algorithms to sift through the information, create

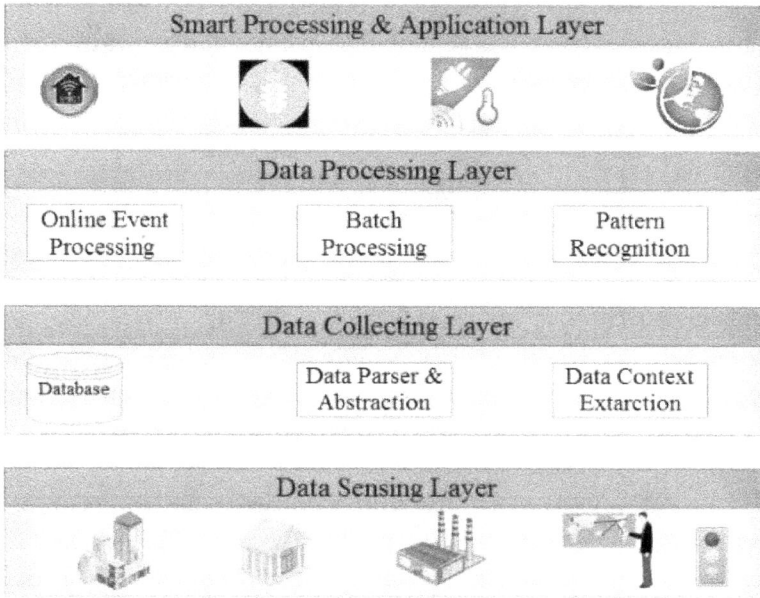

Figure 2. Operational architecture of smart cities.

meaningful context from it, and deal with the problem's massive scale.

III. **Data processing layer:** Smart city applications require per-processing of data collected in locally or distant databases. This layer handles a lot of data; hence, real-time computing and batch processing are crucial. However, it is hard to handle and maintain the obtained large data utilizing algorithms, since they are only relevant to regular data with restricted and well-defined datasets. Consequently, it is vital to create complicated and advanced strategies that are able to analyze massive amounts of data with a high degree of diversity in a dynamic and diverse context. Furthermore, the per-processing step can be executed at the network's periphery, which can minimize latency in delay-sensitive applications (fog computing).

IV. **Integrated processing & application layer:** This layer is one of the primary components of the smart city design. This layer has the capability to carry out precise data analysis, which may then be used to arrive at global decisions and provide raw data to apps for smart cities.

2.   Security Need of Smart Cities

By the year 2015, there was to be the potential for around 15 billion devices to be linked to the Internet. By the year 2020, this number was expected to increase to approximately 50 billion [7]. It also has an effect on practically everything, ranging from people's personal life and education to their health and the safety of their nation. Smart cities make life simpler and aid in managing and controlling many environmental factors. The diversity, interrelations, and interconnectivity of smart cities make them more susceptible to security and privacy breaches. A smart city may not be implemented and operated in a secure manner if the security requirements and problems are not understood.

2.1.   *Secure communication*

In order to connect the many components for data collection, exchange, and transfer, communication is a crucial component of smart city designs. Confidentiality, integrity, and non-repudiation are essential components in establishing secure wired and wireless communication in smart cities. The next obstacle to securing communication is creating a reliable key management system [8].

A CryptoManager IoT Device Management solution was recently presented by Rambus [9] to identify and authenticate IoT devices based on pre-provisioned unique device keys. Using this management system, IoT gadgets may set up encrypted connections to other gadgets and services. When an IoT device connects to a network like the Internet, it has to be verified without causing any disruptions in network traffic. The solution also produces the IoT device's necessary security credentials. However, a large range of IoT devices cannot be supported by this crypto scheme.

2.2. Secure booting

Preboot malware uses an OS kernel to operate before the machine is managed and hides from OS and virus scanners. When a machine boots up using infected media or over an Internet connection, the virus code can be found as executable code and transferred to other computers.

The term "secure booting" refers to a method that assists the system firmware in determining whether or not a cryptographic signature for the system boot loader exists. Since the signature key is kept in the firmware, it is challenging for the person who created the virus to sign it if the key is under the possession of an authorized user. A technique called secure boot is essential for the operation of smart city devices because it ensures the completeness and authenticity of software packages and prevents the signing of unsigned code [10].

A majority of boot-securing solutions are not suitable for IoT devices, despite the fact that secure booting is necessary for IoT devices to safeguard the authenticity and integrity of the software on such devices based on cryptographic hash algorithms. This is due to the minimal processing resources that such devices have. In order to develop an effective boot security for IoT devices, ultra-low-power consumption hash functions must be used. Examples of hash functions for extremely low-power consumption devices are the NH and WH universal hash functions [11,12].

2.3. Security monitoring and analysis

All IoT systems must have a monitoring method in order to govern their environments, identify active threats, and track anonymous activity. Due to the scale of the IoT network, automated reply frameworks must have access to sufficient data about threats and automatic detection of illegal behavior. Various techniques to countering attacks and fraudulent activity could be taken into account by the system.

Disclosure of possible vulnerabilities in smart city systems is necessary for the prompt migration, modification, and upgrading of all impacted parties. This is due to the fact that the IoT devices, which are tasked with the responsibility of gathering and distributing data, are susceptible to a broad range of attack types [13].

For example, Cisco created a security monitoring and analysis system to detect network vulnerabilities by analyzing the network architecture and setup, provide suggestions for lowering vulnerabilities by determining the cause of threats, and handle incidents. This method, therefore, is only usable with Cisco networking equipment [14].

2.4. *Security solution life cycle management*

Smart cities utilize IoT devices to gather, evaluate, and communicate with residents. The need to link all different kinds of gadgets compels engineers to ensure the safety of such connections. Protective measures must be considered at all levels, including the device, the network, and the cloud, while developing software. Managing the IoT system's life cycle requires a great deal of complexity.

In [15], a new data management life cycle model called Smart City Comprehensive Data Life Cycle (SCC-DLC) was proposed for smart cities based on fog and cloud resource management architecture. The SCC-DLC model efficiently manages and organizes large amounts of data from varied information sources throughout smart city data collecting, processing, and storage. This study's primary contribution is adding a new layer to the smart city design that contains a number of fog nodes to increase computation and storage capacity while lowering network traffic and communication latencies. The security and privacy of data have not been quantified in this design, even though decreasing data transmission latency may have a beneficial influence on security risk and communication failure.

2.5. *Security update and patching*

Security update and patching are critical requirements for IoT devices to function effectively and remain safe from the latest harmful threats. Additionally, the devices must be capable of verifying the authenticity of downloaded fixes with their service providers and operators. Authentication methods for IoT devices must not compromise the integrity of the devices' communications or compromise the devices' ability to perform their intended functions.

Even though security patching and updating are the best ways to avoid vulnerabilities, security upgrading or patching is one of the major hurdles for IoT devices. For these devices with medical applications, it is not feasible to install and operate any antivirus as a third-party security approach. Two arguments support this assertion. The first argument is that many medical device manufacturers lack the necessary expertise to facilitate dynamic patch updates, leading to a reliance on secure communication channels for transmitting the collected data. The second argument is that the regulatory restrictions imposed by the Food and Drug Administration make upgrading medical equipment a cumbersome and time-consuming process.

2.6. *Authentication and access control*

The ability of IoT systems and devices to function depends heavily on data exchange. Controlling and managing the data produced by other IoT devices is crucial, as is avoiding its usage in unlawful or undesirable ways. Data security and privacy have been protected by a variety of authentication and access control systems [17].

2.7. *Data and application protection*

Smart cities place a high priority on data security and privacy practices. Smart cities must use various techniques to pinpoint system weaknesses and offer several degrees of data security against internal and external attacks. Using current methods to secure IoT devices is the initial smart city data security stage. It is necessary to protect the confidentiality of the information stored in smartphone applications. The security depends on two things: (i) Protecting the application from disclosing the device's unique identifiers (such as its UDID, IMEI, and MEID) and (ii) keeping tabs on the permissions granted to apps so they can gain entry to sensitive information [18]. Then, current cryptographic techniques and key management methods must be used to secure data transmission between smart city components

Table 1. Impact of vulnerabilities on smart city infrastructure

Requirements	Method	Challenge
Secure Communication	Lightweight cryptographic Methods	Heterogeneity of Network components and devices
	Distributed key management system	Geographical distribution of smart cities; Draining the embedded system's resources
Secure Booting	Cryptographic boot system	Adoption to heterogeneous IoT devices
Security Monitoring, Analysis, and Response	Cisco Security Monitoring, Analysis, and Response System (MARS)	Only applicable for Cisco network equipment
System, Application, and Solution Lifecycle Management	Smart City Comprehensive Data Life Cycle model	Lack of security and privacy measurement
Updating and Patching	Microsoft and Linux patch updating	Authenticating the update package may reduce the IoT device functionalities; May not be applicable for old IoT devices
Authentication, Identification, and Access Control	IBE ABE RBAC	Are only applicable for cloud-based IoT systems; May incur high computation cost on IoT devices
Data and Application Protection	Securing IoT devices, Access permission monitoring, Securing communication links using cryptographic methods	Lack of a comprehensive framework to provide security and privacy of all layers of smart cities simultaneously

and safeguard end-to-end and point-to-point connections. To avoid data leakage and inappropriate use, the various data formats stored in databases and even at the whole disc level must be encrypted [19].

Table 1 analyzes typical needs for smart city security based on approach and challenge features. The method attribute details the many approaches that can be taken to fulfill the necessary safety standards for smart cities.

3.   Vulnerability of Smart Cities

Attackers can utilize the architecture of smart cities to construct and install self-propagating malware that can spread over several interconnected networks. Intruders can readily access users' medical records, bank account details, and other critical information.

They can also perform various cyber threats to destroy (i) confidentiality to obtain information and supervise system activities, such as illegal data gathering through snooping or analyzing message traffic; (ii) integrity to modify information and change system settings, such as unauthorized access to sensitive information; and (iii) availability to close and make the system unavailable for authorization. Table 2 examines a few of the recent intentional attacks and illustrates how they affect smart cities.

Distributed Denial-of-Service (DDoS) attacks can have catastrophic effects on various elements of smart cities, including traffic and security cameras [20–22]. Massive amounts of data are created on the Web in a phenomenon known as DDoS, which is used to disrupt network services and the services offered by smart cities. Data centers are frequently needed in smart cities to house a variety of Web-based application servers, which are frequently the target of such attacks. However, it is highly challenging to suggest a practical way to resist such attacks due to the unique and scattered structure of smart cities in terms of network volume of traffic, velocity, and diversity. Further difficulties include the heterogeneity of networks, high availability, and scalability, as well as the need to take into account the evolving security rules of smart cities when building a mitigation strategy.

In order to resolve these problems, Bawany *et al.* [23] suggested an efficient framework to recognize and counteract the flash crowd application layer DDOS attack in smart cities, which occurs when the number of legitimate connections to a server or website suddenly increases at the same time or over a short period of time. According to the Software Defined Networking (SDN) concept, the basic idea behind this technology is to use a master city controller component

Table 2. Vulnerability and their effects on smart cities

Attack	Key Features
Eavesdropping	• Capturing network traffic and listening to communications between two or more parties. • Disclosing details regarding the configuration of the network.
Cross-Site Request Forgery (CSRF)	Forcing an end user to execute unwanted actions on an attacker web application to perform state-changing requests, such as transferring funds and compromise the whole web application.
SQL Injection Attack	Inserting a SQL query via the input data from the client to the application to read or modify data, or execute administration operations on the database.
Cross Site Scripting (XSS)	Injecting client-side scripts into web pages by attacker to evade access controls such as the same-origin policy.
Side-Channel Attack	Exploiting the available information (e.g., plaintext, cyphertext, or timing information) to find a user's key and retrieve data from a encrypted device.
Distributed Denial of Service (DDoS)	• Overloading a targeted resource by consuming available bandwidth. • Overwhelming targeted resources by using protocol flaws. • Overloading application services or databases with a high volume of application calls.
Brute-Force Attack	Using many passwords or pass phrases to eventually guess the password and hack into network.
Replay Attack	Eavesdropping a stream of messages between two parties and fraudulently retransmit it to one party to perform unauthorized operation, such as false identification and authentication.
Session Hijacking	Exploiting a valid session key or stealing a magic cookie of an authorized user to acquiring unauthorized access to information or services.
Virtual Machine (VM) Escape	Breaking out a virtual machines (VM) and interacting directly with the hypervisor to obtain access to the host operating system and other VMs running on that host.
Unauthorized Access	Including unauthorized network connection, data leaks, browsing files, obtaining private data, controlling field components and using resources.

to efficiently monitor and separate harmful traffic from legitimate traffic flows [34].

A side-channel attack is a type of physical attack that can be used against smart cities. In this type of attack, the adversary compromises the safety of the majority of cryptographic IoT devices by utilizing information leakages that are caused by the device's physical properties, such as its power consumption and timing information. Brute-force attack, in which an attacker uses a known flaw in the network to break into the system and steal sensitive information, is another sort of security threat in smart cities.

4. Open Research Issues and Challenges

Smart cities have just evolved as a new paradigm, and there are a number of investigators working on various parts of the smart city. Although security is considered the most important part of smart cities, it has not been properly considered by researchers. This section lists open issues for future consideration.

4.1. *Lightweight security methods for smart city*

Sensors dispersed across cities capture a vast quantity of data in a variety of forms from many facets of urban life, including the transportation system, residential areas, educational institutions, manufacturing facilities, and medical facilities. Processing either in real time or in batches is done of the gathered data in order to detect events as they occur. Such a large volume of data demands a very effective and lightweight cryptographic method to maintain security and privacy [24,31,33].

4.2. *Outsourcing of secure data*

Cloud-based frameworks solve the smart city storage overhead efficiently. Even though the cost of storage hardware is decreasing, it is getting harder to manage such large amounts of storage, which makes up about 75% of the total cost of ownership [25,32]. Thus, smart cities' massive IoT data may be outsourced to cloud storage and managed by CSPs. However, the cloud could be more reliable and poses additional threats to outsourced data in cloud computing. To deal with this problem, Remote Data Auditing (RDA) techniques were devised, wherein the data owner may verify the authenticity of the outsourced data without actually downloading it. However, there are several RDA techniques to verify the accuracy of data that have been outsourced; these techniques are not suitable for supporting big data through smart cities.

4.3.   *Secure and adaptive sensing for smart cities*

Participants utilize mobile phones and cloud services to gather and assess systematic data for discovery. Collaborative sensing is able to supply the nearby sign and information on environmental factors obtained by end users, which together create a social currency. This is another one of the capabilities of collaborative sensing. Thus, smart city applications like healthcare and energy control may immediately compare existing data with online data across a fixed infrastructure sensor network. These applications can also leverage end user input on an environmental parameter. However, smart cities' infrastructure hinders them from using these capabilities; therefore, they need a new framework to harness the immense potential of participatory sensing to gather data from trustworthy sources and execute proper analysis [26].

4.4.   *Security risk management and mitigation*

The sensor technology used in smart cities is typically installed in difficult locations with multiple security risks. Designing a security strategy to reduce such vulnerabilities is crucial for enabling cloud-based smart city applications. Most apps migrate from physical servers to virtual servers without taking data security into account [27].

4.5. *Fog computing in smart cities*

IoT devices typically capture massive amounts of data, making it difficult for smart cities to store and handle such massive amounts of data. To address this issue, IBM has suggested fog computing, a new idea in which data may be processed at the network's edge rather than being sent to the cloud.

The use of fog computing in the construction of a sustainable smart city is very applicable. Fog can be used to examine the data gathered by sensors and GPS units that have been placed in the field. There are various case studies that demonstrate that fog computing plays an important role in the management of both water and agricultural resources. While fog computing has the potential to greatly improve smart cities in many ways, it also introduces new virtualization, Web, and data security difficulties. This is due to the fact that fog nodes have limited access to computing resources, which makes it challenging to provide solutions to security issues [28]. The fact that fog nodes are frequently easier to reach than cloud data centers raises the likelihood of intrusions. Fog nodes are even more appealing to attackers since they gather massive amounts of valuable information from various sources [29].

4.6. *Implementing effective government policies*

In smart cities, it is up to IoT devices to collect data and send them to local or remote data centers for batch or real-time data analysis. Smart cities could also sell the information they gather to third-party companies without asking the people who live there first. Some cities, such as New York, London, and Dubai, have recently installed new surveillance cameras in order to record criminal activity and terrorist strikes. These devices endanger the safety of smart cities because they violate residents' civil rights by tracking their travels and permitting arbitrary searches [30].

5. Conclusion

In this chapter, we have classified and categorized a wide range of research fields connected to the smart city, with the goal of focusing on vulnerability issues and ways to address them. Specifically, we are interested in focusing on how to handle the difficulties. We started off by describing the notion of smart cities and contrasting their various characteristics. Then, we provided classifications of the smart city based on its architecture and its uses. Additionally, we focused on the challenges, issues, and requirements of security for smart cities. Further, we critically assessed and looked at the potential remedies for resolving the security problems in smart cities. Finally, several open issues that are major hurdles for future research were mentioned.

Acknowledgment

The authors would like to express their gratitude to the reviewers whose careful reading and insightful remarks significantly enhanced the quality of this chapter.

References

[1] Gomez, A., Shahriar, H., Clincy, V., and Shalan, A. Hands-on lab on smart city vulnerability exploitation. In *2020 IEEE 44th Annual Computers, Software, and Applications Conference (COMPSAC)*, pp. 1777–1782. IEEE, 2020.

[2] Maruf, M.H., *et al.* Adaptation for sustainable implementation of Smart Grid in developing countries like Bangladesh. *Energy Reports*, 6:2520–2530, 2020.

[3] Chen, Z. Application of environmental ecological strategy in smart city space architecture planning. *Environmental Technology & Innovation*, 23:101684, 2021.

[4] Lebrument, N., *et al.* Triggering participation in smart cities: Political efficacy, public administration satisfaction and sense of belonging as drivers of citizens' intention. *Technological Forecasting and Social Change*, 171:120938, 2021.

[5] Cerrudo, C. An emerging us (and world) threat: Cities wide open to cyberattacks. *Securing Smart Cities*, 17:137–151, 2015.

[6] Al Dakheel, J., *et al.* Smart buildings features and key performance indicators: A review. *Sustainable Cities and Society*, 61:102328, 2020.

[7] Nordrum, A. Popular Internet of Things forecast of 50 billion devices by 2020 is outdated. IEEE Spectrum, New York, NY, USA, Rep., 2016.

[8] Shi, Q., Xu, W., Wu, J., Song, E., and Wang, Y. Secure beamforming for MIMO broadcasting with wireless information and power transfer. *IEEE Transactions on Wireless Communications*, 14(5):2841–2853, 2015.

[9] Cyber security in the era of the smart home. Rambus Inc., Sunnyvale, CA, USA, Rep., 2016 [Online]. Available at: http://info.rambus.com/hubfs/ rambus.com/Gated-Content/Cryptography/Cyber-Security-in-the-Era-of- the-Smart-Home-White-Paper.pdf?hsCtaTracking=bf8d7eb4-8b77-4a07-8a e2-759100c676a6%7C69c7a89b-6dfb-4926-8c38-860b7fd1eff5.

[10] Lebedev, I., Hogan, K., and Devadas, S. Secure boot and remote attestation in the sanctum processor. In *2018 IEEE 31st Computer Security Foundations Symposium (CSF)*. IEEE, 2018.

[11] Sengupta, J., Ruj, S., and Bit, S.D. A comprehensive survey on attacks, security issues and blockchain solutions for IoT and IIoT. *Journal of Network and Computer Applications*, 149:102481, 2020.

[12] Hassan, M.U., Rehmani, M.H., and Chen, J. Differential privacy techniques for cyber physical systems: A survey. *IEEE Communications Surveys & Tutorials*, 22(1):746–789, 2019.

[13] Sookhak, M., *et al.* Security and privacy of smart cities: A survey, research issues and challenges. *IEEE Communications Surveys & Tutorials*, 21(2):1718–1743, 2018.

[14] Cisco Security Monitoring, Analysis and Response System [Online]. Available at: https://www.cisco.com/c/en/us/products/security/securitymonitoring- analysis-response-system/index.html.

[15] Sinaeepourfard, A., Garcia, J., Masip-Bruin, X., and Marin-Tordera, E. A novel architecture for efficient fog to cloud data management in smart cities. In *Proceedings of the IEEE 37th International Conference on Distributed Computing Systems (ICDCS)*, pp. 2622–2623, June 2017.

[16] Zou, X. IoT devices are hard to patch: Here's why and how to deal with security, Techbeacon [Online]. Available at: https://techbeacon.com/ iot-devices-are-hard-patch-heres-why-howdeal-security.

[17] Liu, J., Xiao, Y., and Chen, C.L.P. Authentication and access control in the Internet of Things. In *2012 32nd International Conference on Distributed Computing Systems Workshops*. IEEE, 2012.

[18] Zang, J., Dummit, K., Graves, J., Lisker, P., and Sweeney, L. Who knows what about me? A survey of behind the scenes personal data sharing to third parties by mobile apps. Technology Science, Rep., October 2015.

[19] Bahrami, M. and Singhal, M. The role of cloud computing architecture in big data. *Information Granularity, Big Data, and Computational Intelligence*, pp. 275–295. Springer, Cham, 2015.

[20] Bhale, P., *et al.* Energy efficient approach to detect sinkhole attack using roving IDS in 6LoWPAN network. In *International Conference on Innovations for Community Services*. Springer, Cham, 2020.

[21] Bhale, P., Biswas, S., and Nandi, S. LORD: LOw Rate DDoS attack detection and mitigation using lightweight distributed packet inspection agent in

IoT ecosystem. In *2019 IEEE International Conference on Advanced Networks and Telecommunications Systems (ANTS)*. IEEE, 2019.

[22] Bhale, P., Biswas, S., and Nandi, S. An adaptive and lightweight solution to detect mixed rate IP spoofed DDoS attack in IoT ecosystem. In *2018 15th IEEE India Council International Conference (INDICON)*. IEEE, 2018.

[23] Bawany, N.Z., Shamsi, J.A., and Salah, K. DDoS attack detection and mitigation using SDN: Methods, practices, and solutions. *Arabian Journal for Science and Engineering*, 42(2):425–441, 2017.

[24] Zhang, W., *et al.* LDC: A lightweight dada consensus algorithm based on the blockchain for the industrial Internet of Things for smart city applications. *Future Generation Computer Systems*, 108:574–582, 2020.

[25] Sookhak, M., Talebian, H., Ahmed, E., Gani, A., and Khan, M. A review on remote data auditing in single cloud server: Taxonomy and open issues. *Journal of Network and Computer Applications*, 43:121–141, 2014.

[26] Shi, Q., Xu, W., Wu, J., Song, E., and Wang, Y. Secure beamforming for MIMO broadcasting with wireless information and power transfer. *IEEE Transactions on Wireless Communications*, 14(5):2841–2853, 2015.

[27] Meland, P.H., Tondel, I.A., and Solhaug, B. Mitigating risk with cyberinsurance. *IEEE Security & Privacy*, 13(6):38–43, 2015.

[28] Sun, P.J. Security and privacy protection in cloud computing: Discussions and challenges. *Journal of Network and Computer Applications*, 160:102642, 2020.

[29] Gill, S.S., *et al.* Transformative effects of IoT, Blockchain and Artificial Intelligence on cloud computing: Evolution, vision, trends and open challenges. *Internet of Things*, 8:100118, 2019.

[30] Maestre, G., *et al.* Empirical evidence of Colombian national e-government programs' impact on local Smart City-Adoption. In *Proceedings of the 11th International Conference on Theory and Practice of Electronic Governance*, 2018.

[31] Bhale, P., *et al.* OPTIMIST: Lightweight and transparent IDS with optimum placement strategy to mitigate mixed-rate DDoS attacks in IoT networks. *IEEE Internet of Things Journal*, 10(10):8357–8370, 15 May 2023, doi: 10.1109/JIOT.2023.3234530.

[32] Dey, S., *et al.* ReFIT: Reliability challenges and failure rate mitigation techniques for IoT systems. In *International Conference on Innovations for Community Services*. Springer, Cham, 2020.

[33] Das, A., *et al.* A novel approach to detect rank attack in IoT ecosystem. In *International Conference on Innovations in Intelligent Computing and Communications*. Springer, Cham, 2022.

[34] Ray, D., *et al.* DAISS: Design of an attacker identification scheme in CoAP request/response spoofing. In *TENCON 2021–2021 IEEE Region 10 Conference (TENCON)*. IEEE, 2021.

Part 3
Limitations

Chapter 6

Cascading Failures in Smart Grids under Random, Targeted, and Adaptive Attacks

Sushmita Ruj[*,‡] and **Arindam Pal**[†,§]

UNSW Sydney, New South Wales, Australia
†*Data61, CSIRO, Sydney, New South Wales, Australia*

‡*sushmita.ruj@gmail.com*
§*arindamp@gmail.com*

Abstract: A smart grid consists of two networks: the power network and the communication network, which are interconnected by edges spanning across the networks. We model smart grids as complex interdependent networks, and study targeted and adaptive attacks on smart grids for the first time. Due to an attack on one network, nodes in the other network might get isolated, which in turn will disconnect nodes in the first network. Such cascading failures can result in disintegration of either or both of the networks. Earlier works considered only random failures. In real life, an attacker is more likely to compromise nodes selectively.

We study cascading failures in smart grids, where an attacker selectively compromises the nodes with probabilities proportional to their degrees, betweenness, or clustering coefficient. We mathematically and experimentally analyze the sizes of the giant components of the networks under different types of targeted attacks, and compare the results with the corresponding sizes under random attacks. We show that networks disintegrate faster for targeted attacks compared to random attacks. We next study adaptive attacks, where an attacker compromises nodes in rounds. Here, some nodes are compromised in each round based on their degree, betweenness or clustering coefficients, instead of compromising all nodes together. We show experimentally that an adversary has an advantage in this adaptive approach, compared to compromising the same number of nodes all at once.

Keywords: Complex Networks, Percolation Theory, Smart Grids, Cascading Failures, Degree, Betweenness, Clustering Coefficients, Random, Targeted and Adaptive Attacks.

1. Introduction

Power grids have suffered severe failures in the past. The blackout of Northern US/Canada and that of Italy in 2003 affected the lives of millions of people and resulted in huge monetary losses. More recently, the largest blackout in the world occurred in India in July 2012. The complete shutdown of the Northern, Eastern, and Northeastern power grids in the country affected over 620 million people. Such calamities could have been avoided if the power grid functioned properly. In order to ensure that the electric grid functions smoothly, it is important that the control information is collected and transmitted in an orderly fashion, and the existing systems are highly automated. Smart grids are next-generation electricity grids, in which the power network and the communication network work in tandem. Smart grids promise to fulfill this vision by synchronizing the power network with the communication network. The idea is to replace the existing *SCADA (Supervisory Control and Data Acquisition)* system by an intelligent and automatic communication network.

The power network consists of power plants and generation and distribution stations, whereas the communication network consists of sensors attached to appliances to collect information, aggregator sensors to aggregate information, and smart meters for monitoring and billing. The smart meters in home area networks, building area networks, and neighborhood area networks are responsible for aggregating, processing, and transmitting data and control information for proper functioning of the smart grid. The question is how to make such a network robust and fault tolerant. Researchers have addressed smart grid architectures [4] and the problem of cascading failures [11], in which a small fault propagates throughout the network and affects a large part of the network. Most of the current techniques and models use concepts from distributed systems. However, because of the large size of smart grids and their unique properties, new models, interconnection patterns, and analysis techniques are required to increase the robustness of networks.

Recently, Huang *et al.* [18] initiated the study of modeling and analyzing smart grids using interdependent complex networks. A smart grid can be thought of as two complex networks, which are interconnected. The question is how to make this network robust and fault tolerant. In order to provide a solution, we have to understand what kind of faults and attacks can take place and how faults propagate in the network. The failure of nodes in one network results in the disruption of the other network, which in turn affects the first network. This type of failure propagates in a cascading manner and was the main reason for the blackouts in the US and in India. To understand this *cascading failure*, we need to study the structure of the networks. In this chapter, we model and study smart grids as complex networks and show the effect of cascading failure, when adversaries compromise nodes in the network.

Though cyber-security issues have been studied in detail [49], modeling the network in order to make it resilient still needs a lot of research. The main contribution of this chapter is to study the effect of targeted and adaptive attacks on smart grids, in which the attacker selectively disrupts communication nodes. This is one of the early works on targeted and adaptive attacks on smart grids using the complex network model. We argue that an adversary is more likely to attack selected high-degree (or betweenness or clustering coefficient) nodes, rather than attacking nodes randomly. As an example, we consider the recent Stuxnet worm [26] which targeted Siemens PCs and caused large-scale destruction to industrial control systems. Yagan *et al.* [53] studied cascading failures in cyber-physical systems. They studied different interdependent Erdos–Renyi (ER) networks [37], but they did not consider scale-free (SF) networks, which are used to represent power and communication networks. Till date, all works [18,19,44,53] on complex network models of the smart grid have considered only random attacks. In the preliminary version [44] of this paper, we addressed targeted attacks. We analyzed the sizes of giant components in each network under targeted attacks. However, nodes were compromised based only on the degree of the node, i.e., high-degree nodes were compromised with high probability. Betweenness and clustering coefficients were not considered. Adaptive attacks were also not considered in the preliminary version. Huang *et al.* [19] addressed the cost of maintaining such networks by analyzing the number of support links between networks. Whereas increasing the support links might make the interdependent networks stronger, the large number of support links implies higher cost of maintenance. They suggested that smart grids should have some nodes which are connected to power nodes (also called *operation centers*) and the rest of the nodes are *relaying nodes*. Using such a model, they studied the resilience of the network under random attacks. According to their model, each control node is linked to n power nodes and each power node is operated by k operation centers.

Interdependent networks have been studied in the context of cyber-physical systems in general by [20]. They studied cascading failures in interdependent networks. The paper studies random attacks on the networks. They calculate percolation thresholds for interdependent networks using extensive experiments. Interdependent networks have also been addressed for complex contagion in [7].

1.1. *Problem statement and our contribution*

We model the smart grid as a complex interdependent network consisting of two networks, the power network and the communication network. Both the power network and the communication network are SF networks, where the degree distribution follows the power law, $p_k \propto k^{-\alpha}$, where p_k is the fraction of nodes of degree k and α is the power law parameter specific to the network. SF networks are a type of random graph which commonly arises in many practical cases, like social, biological networks, the Internet, and power grids. Another type of network, which is often studied is the ER network, denoted by $G(n, p)$, where n is the number of nodes and p is the probability that an edge exists between two nodes. Support links are randomly assigned from one network to another, such that a power node is controlled by multiple communication nodes and functions properly as long as at least one such link exists. In our model, we consider targeted attacks on the communication network. We mathematically analyze the effect of cascading failure for this type of attack and find out the sizes of giant components when nodes are compromised.

We compare the following attack models: random attacks, targeted attacks, and a combination of targeted and random attacks. We show that an adversary has a definite advantage if it compromises nodes selectively. A simple example is that if an adversary wants to launch a terrorist attack, it would like to plant as few bombs as possible, while maximizing the damage. Thus, the adversary has to compromise nodes selectively. Critical node detection is an interesting problem, which has been studied in literature in many contexts. As pointed out in [38,46], detecting critical nodes in an interdependent power grid is an NP-complete problem.

In this chapter, we have considered the following strategies for compromising nodes selectively: the attacker might consider the degree, betweenness, or clustering coefficient and compromise nodes with high degree (or betweenness or clustering coefficient). *Betweenness* of a node is the number of shortest paths that pass through a node. *Clustering coefficient* is the number of common neighbors of two nodes which are neighbors themselves. A formal definition appears in Section 3. In targeted attack, the adversary compromises a node with a probability proportional to the degree (or betweenness or clustering coefficient) of the node. We show that, from the point of view of the adversary, compromising nodes with probability proportional to the betweenness is better than compromising nodes either randomly or with probability proportional to the degree or clustering coefficients. We compare our results for a combination of SF–SF networks and ER–ER networks. We show that SF–SF networks are more vulnerable to targeted attacks than ER–ER networks. We also analyze the average path length under targeted attacks.

Next, we compromise nodes adaptively. This means that instead of selecting all nodes to be compromised at the start, we compromise nodes (based on degree/betweenness/clustering coefficients) in rounds. This implies that an attacker compromises a few nodes in round one and then, depending upon the cascading effect on the two networks, compromises another set of nodes in the second round and so on. In this case, we show that an adversary has an advantage for adaptive attacks over non-adaptive attacks. Adaptive attacks based on betweenness result in smaller giant components compared to Adaptive attacks based on degree.

Our main conclusion is that by launching a targeted attack, an adversary can disrupt a significant part of the network. For a large network, compromising about 2.2% of the network can disrupt either of the networks under targeted attack, whereas under random attack, the networks are still connected and work smoothly.

We observe that after targeted attacks, the size of the giant component in ER networks can be twice as large as that in SF networks.

1.2. *Organization*

The chapter is organized as follows. Related works are presented in Section 2. Preliminary material on complex networks is given in Section 3. Network model and attack model are presented in Section 4. The basic technique for computing the size of the giant component is described in Section 5. Cascading failure is mathematically analyzed in Section 6. In Section 7, we present experimental results to understand our model and make some conclusions. We conclude in Section 8 with directions for future work.

2. Related Works

Smart grid communication and network architecture have been widely studied in [4,29,50]. Most smart grid literature concentrates on distribution of power [55], balancing supply and demand [41], detecting and predicting faults [11], and designing network architecture which is fault tolerant [53]. The bulk of the literature on fault tolerance addresses cyber-physical systems in general [53] and uses general models and techniques of distributed systems.

The cybersecurity requirements of smart grids have been outlined by the National Institute of Standards and Technology (NIST) [27]. There has been extensive research on the security and privacy of smart grids in recent years. Surveys can be found in [29,49]. The main problems that have been explored include secure and privacy-preserving data aggregation in smart meters which have been studied in [28,33,42,43]. Privacy-preserving smart metering has been addressed in [1,36]. Access control of smart grid data has been presented in [3,43]. Data authentication has also been studied by Fouda *et al.* [15], Kgwadi-Kunz [25], and Lu *et al.* [33,34].

A coordinated multi-switch attack was proposed in [30]. The opponent is able to control many switches in the power system. Using dynamical systems, the authors show how to launch an attack by using multiple beakers. Other switching attacks have been studied in [31,32].

Fault tolerance in power grids has been studied widely in the past. Efficient methods of load distribution to prevent cascading failures have been studied in [10]. Load Redistribution (LR) attack was developed and studied in [54] by analyzing the extent of damage to power grid operation. The power grid has been modeled as a graph and the robustness has been studied in [22]. By extensive mathematical analysis, the disturbance levels the system can accept before a few overload nodes resulting a large blackout has been estimated. Static overload failure was discussed in [11]. Optimization techniques are used and a distance-to-failure algorithm was proposed to predict the weak points in the power grid. They proposed an attack model describing the main goal of the LR attack, and then based on that

indicated the theory and criterion of protecting the system from the
LR attack. To decrease the impact of overload cascading, [40] proposed a model to focus on the analysis of tripping of already overloaded lines. By simulation on a real-world power grid structure, it is
shown that controlled tripping of overloaded lines leads to significant
mitigation of cascading failure.

The study of the model, the analysis of the network structure, and
increasing the robustness of power grids have been conducted using
complex networks. Here, electric distribution stations, transmission
stations, and generation centers are modeled as nodes. Two nodes
are connected by a link if there is power flow from one node to the
other. The structure of the underlying graph has been widely studied
to find the effect of node failures. When certain nodes fail (or are
attacked), the links incident on these nodes are disrupted. This affects
other nodes, whose links fail in turn. Such failures propagate in a
cascading manner throughout the network. Thus, a small fraction of
nodes can disrupt a large part of the network. It has been shown
that the graph structure underlying a power grid follows a power law
distribution [37]. An extensive survey appears in [39].

Although complex networks have been widely used to study different networks like social networks, biological networks, citation net-
works, and power networks [37], smart grid networks have not been
widely studied. Huang *et al.* [18] introduced the study of smart grids
using complex interdependent networks, in which the power network
and the communication network are modeled as individual networks
which have SF property. The links connecting nodes within a network are called *intralinks*. The networks are connected to each other
via links (also called *interlinks*), such that a power node depends on
communication nodes and vice versa. Such a network is called an
interdependent network.

Interdependent networks were introduced by Buldyrev *et al.* [8].
They studied the effect of failure cascades in such networks. The failure of a few nodes in the communication network will affect nodes
in the power network, which will further affect nodes in the communication network. Thus, failures propagate in cascades till a steady

state is reached or when either or both of the networks disintegrate. We say that a network *disintegrates* if there are no *giant components* in the network. A giant component is a connected component of size $\Theta(N)$, where N is the number of nodes in the network. Since then, a number of researchers have analyzed interdependent networks.

The initial analysis by Buldyrev *et al.* [8] studied the case where the two networks are of the same size, and there is a one-to-one correspondence between nodes which are joined by an interlink. Shao *et al.* [47] studied multiple support interlinks, where a node in the power network was connected to multiple nodes in the communication network and vice versa. Most of the results have been analyzed experimentally, because closed-form analytical solutions are difficult to obtain. A special case of support links, where nodes having identical degree are connected across networks, was studied in [9]. It has been observed in all these cases that interdependent systems make the network much more vulnerable to attacks, compared to a single network.

A well-known result in complex networks is that randomly removing 95% of the nodes in the Internet (which is an SF network) can still result in a connected network. However, strategically removing even 2.5% of the nodes can disrupt the whole network [14]. Such a result motivates us to study the effect of targeted attacks on smart grids. In the case of smart grids, an adversary is more likely to compromise nodes of strategic importance, like hubs, than nodes of low degree. Thus, selective attacks give substantially different results compared to random attacks.

Targeted attacks on interdependent networks have been studied in [17]. The attacker chooses the nodes with probability proportional to the degree. It follows that a high-degree node has a higher probability of being attacked. Our work is significantly different from theirs in the following respects: (i) In [17], the authors considered only targeted attacks based on degree. We have compared targeted attacks based on degree, betweenness centrality, and clustering coefficients, and also random attacks. (ii) The paper [17] assumed that the two networks are of the same size and same type (both ER–ER or both

SF–SF). So, there is one–one correspondence between the nodes in either part of the network. We have considered general networks even with unequal number of nodes in the two parts. (iii) The main aim of [17] was to find the percolation threshold. In this chapter, we compare different parameters like size of the giant component, average path length, and other parameters under different attack strategies, when a given number of nodes are compromised. (iv) Adaptive attack has been studied in this chapter, but not in [17].

Instead of studying interdependent networks consisting of two networks, Dong *et al.* [13] studied targeted attacks on a network of networks. Zheng and Liu [51] proposed a solution for making a network robust against targeted attacks by suggesting an onion-like structure. Here, high-degree nodes are present toward the center in clusters and low-degree nodes are present in concentric rings depending upon their degree. They analyzed results from power networks. Their technique is however restricted to single networks.

Ruj and Pal [45] discussed different network models of smart grids and their impact on the reliability and availability. They analyzed various techniques to increase the resilience of networks. Zhu *et al.* [56] proposed an analytical method, based on complex networks, to assess the risk of Smart Grid failure due to communication network malfunction, associated with latency and ICT network reliability. The proposed approach is tested on a laboratory-scale communication network. Jiang *et al.* [21] developed an evolutionary computation-based vulnerability analysis framework, which employs particle swarm optimization to search the critical attack sequence. Zuniga *et al.* [57] introduced the application of the Failure Modes and Effects Analysis (FMEA) method in future smart grid systems in order to establish the impact of different failure modes on their performance. They proposed a reliability-based approach that makes use of failure modes of the main components of the power and cyber network to evaluate risk analysis in smart electrical distribution systems. Gupta *et al.* [16] proposed a probabilistic framework of smart grid power network with statistical decision theory to evaluate system performance in steady state, as well as under a dynamical case, and identify the probable critical links which can cause cascade failure. They developed a graphical model using minimum spanning trees to analyze topology and structural connectivity of the IEEE 30 bus system.

3.   Background on Networks

In this section, we will define a few terms related to networks that will be used later. The *degree* of a node is defined as the number of edges that are incident on that node. *Degree distribution* is a random variable X, such that $P(X = d)$ is the fraction of nodes which have degree d. *Centrality* of a node measures its relative importance in the network. It is measured by parameters such as degree, betweenness, and clustering coefficient. *Betweenness* of a node is the number of shortest paths that pass through the node. *Clustering coefficient* is measured by two parameters, local clustering coefficient and global clustering coefficient. Local clustering coefficients, also called transitivity, measure the probability that two neighbors of a vertex are connected. More precisely, this is the ratio of the triangles and connected triples in the graph.

Calculating the betweenness centralities of all the vertices in a graph requires finding the shortest paths between all pairs of vertices on a graph, which takes $\Theta(n^3)$ time with the Floyd–Warshall algorithm, by modifying it to find all shortest paths between two nodes. On a sparse graph, Johnson's algorithm takes $O(n^2 \log n + nm)$ time. On unweighted graphs, calculating betweenness centrality takes $O(nm)$ time using Brandes's algorithm [5]. Here, $n = |V|$ and $m = |E|$ are the number of vertices and edges of the graph $G = (V, E)$, respectively. Existing randomized and parallel algorithms, [2,6,23,35], for calculating centrality measures for large graphs can be used. Through-out the chapter, we assume for simplicity that the adversary has complete knowledge of the network. If complete knowledge is not available, then the adversary might use incremental algorithms [24] to calculate centrality. In such cases, the centrality measures are calculated iteratively, as and when the adversary gradually gains more knowledge of the network.

A *giant component* in a graph on n vertices is a maximal connected component with at least cn vertices, for some constant c. If $c = 0.5$, this means that the giant component should have at least half of the vertices in the graph. A giant component gives a measure of the connectivity of a network. For a power grid, if there exists a giant component containing 80% of the vertices, then these vertices can communicate with each other. If all components are small, the connectivity is very limited. The size of the giant component and its existence become important when nodes in a power/communication network are compromised.

4. Smart Grid Model

We will first discuss the network model and then the attack model.

4.1. *Network model*

We consider two interdependent SF networks, a communication network $N_A = (V_A, \alpha_A)$ and a power network $N_B = (V_B, \alpha_B)$, where $n_A = |V_A|$ and $n_B = |V_B|$ are the number of nodes in the communication and power networks, respectively, and α_A and α_B are the power law coefficients. This implies that N_A has the power law distribution $P_A(k) \propto k^{-\alpha_A}$, which means that the fraction of nodes with degree k is $P_A(k)$. Similarly, N_B has the power law distribution $P_B(k) \propto k^{-\alpha_B}$. We assume that there are more communication nodes than power stations, which implies that $n_A > n_B$.

The interlinks, also called support links [47], are directed edges from one network to the other. We assume that a communication link supports one power station and is powered by one power node, meaning that both the in-degree and out-degree of a communication node are one. A power node is controlled by multiple communication nodes and supplies power to multiple communication nodes, meaning both the in-degree and out-degree of a power node are at least one. Links are assigned randomly from the communication network N_A to the power network N_B.

We have considered a simple model for understanding the dynamics of the networks. In the future, we will extend our model and include other relevant parameters. The papers [8,18–20] have previously considered this simple model. Let $\tilde{k}_A$ denote the support degree of a node in Network A. This implies that there are $\tilde{k}_A$ nodes in N_B that support a node in N_A. Let $\tilde{P}_A(\tilde{k}_A)$ denote the degree distribution of support links from N_B to N_A. $\tilde{P}_B(\tilde{k}_B)$ can be defined analogously. From the structure of the network, $\tilde{k}_A$ is equal to one for all nodes in N_A.

To calculate the degree distribution $\tilde{P}_B(\tilde{k}_B)$, we note that the problem of assigning support links from N_A to N_B is equivalent to assigning n_A balls randomly into n_B bins. If X_i denotes the random variable that counts the number of balls in bin i, then,

$$Pr[X_i = k] = \binom{n_A}{k} \left(\frac{1}{n_B}\right)^k \left(1 - \frac{1}{n_B}\right)^{n_B - k}.$$

Thus, the degree distribution $\tilde{P}_B(\tilde{k}_B)$ follows Binomial distribution with parameters $\mathrm{Bin}\left(n_A, \frac{1}{n_B}\right)$.

4.2. *Attack model*

We consider a targeted attack on communication a network. The attacker chooses the nodes with probability proportional to the degree or betweenness or clustering coefficient. It follows that a high degree/betweenness/clustering coefficient node has a higher probability of being attacked. Targeted attacks are more likely to arise in real-world situations, as we have seen during the recent Stuxnet attack. Attacking the high-degree node is also intuitive, since disrupting the high-degree nodes results in more connections being disrupted, thus disrupting the network.

Table 1. Notations.

N_A	Communication network
N_B	Power network
n_A, n_B	Number of nodes in N_A and N_B
k	Degree of a node
$P_A(k)$	Degree distribution of communication network
$P_B(k)$	Degree distribution of power network
$\tilde{P}_A(\tilde{k}_A)$	Degree distribution of support degree of a node in N_A
$\tilde{P}_B(\tilde{k}_B)$	Degree distribution of support degree of a node in N_B
$\mathcal{G}_{A_n}$	Giant component of N_A at stage n
$\mathcal{G}_{B_n}$	Giant component of N_B at stage n
q_{A_k}	Probability of a node having excess degree k (*i.e.*, total degree $k+1$) in N_A
q_{B_k}	Probability of a node having excess degree k (*i.e.*, total degree $k+1$) in N_B
r_{A_n}	Fraction of removed nodes in N_A at stage n, due to removal of nodes in N_B at stage $n-1$
r_{B_n}	Fraction of removed nodes in N_B at stage n, due to removal of nodes in N_A at stage $n-1$
μ_{A_n}	Fraction of functional nodes in N_A at stage n
μ_{B_n}	Fraction of functional nodes in N_B at stage n

We also consider adaptive attacks which we have discussed in Section 6.7.

A vertex can be deleted from the graph in any of the following cases:

(1) If the vertex is attacked.
(2) If the vertex becomes isolated.
(3) If the vertex is not attacked, but all its support nodes on the other network have been attacked.

Note that due to this kind of cascading failure of nodes, many more nodes will be compromised. This is different from the normal scenario, where only the attacked nodes are compromised.

4.3. *Notations*

We have used the notations given in Table 1 throughout the chapter. Figure 1 shows a smart grid as an interdependent network.

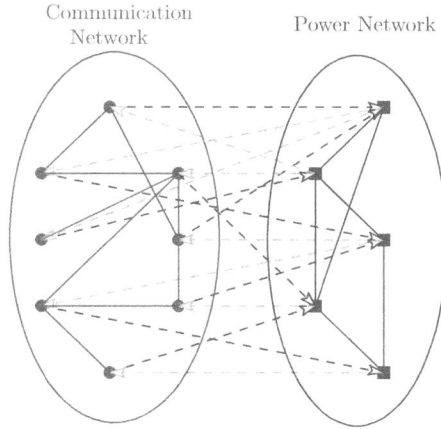

Figure 1. The smart grid as an interdependent complex network.

5. Calculating Giant Component upon Random Removal of Vertices

In this section, we will study the effect of random node compromise on a single network. We will see how this result can be used for analyzing attacks on interdependent networks. Let us consider a network N having a degree distribution $P(k)$. Let ϕ be the fraction of nodes left after random removal of nodes. Let u be the probability that a vertex is not connected to the giant component via a particular neighbor. If the vertex has degree k, then average probability that it is not in the giant component is

$$g_0(u) = \sum_k P(k)u^k, \qquad (1)$$

where $g_0(z) = \sum_k P(k)z^k$, is the generating function for the degree distribution. Hence, the probability that a vertex belongs to a giant component is $1 - g_0(u)$. However, the vertex itself is present with a probability ϕ. Thus, fraction of nodes in the giant component is

$$\mu_N = \phi(1 - g_0(u)). \tag{2}$$

In order to calculate the value of u, we note that a node i is not in the giant component if it is either removed or it is present but not connected to the giant component via any of its neighbors. The first condition happens with probability $1 - \phi$, whereas the second condition happens with probability ϕu^k. Since node i can be reached following an edge, the value of k follows the excess degree distribution

$$q_k = \frac{(k+1)q_k + 1}{\langle k \rangle}, \tag{3}$$

where $\langle k \rangle$ is the average degree of the network. Thus, averaging over this distribution, we get

$$
\begin{aligned}
u &= \sum_{k=0}^{\infty} q_k(1 - \phi + \phi u^k) \\
&= 1 - \phi + \sum_{k=0}^{\infty} q_k u^k \\
&= 1 - \phi + \phi g_1(u),
\end{aligned}
\tag{4}
$$

where

$$g_1(z) = \sum_{k=0}^{\infty} q_k z^k \tag{5}$$

is the generating function for excess degree distribution. A detailed analysis can be found in [37].

6. Modeling Cascading Failure Due to Targeted Attack on Communication Network

We first analyze the targeted attack on a communication network and then show how the failure propagates across the interdependent networks in stages. This is represented in Figure 2.

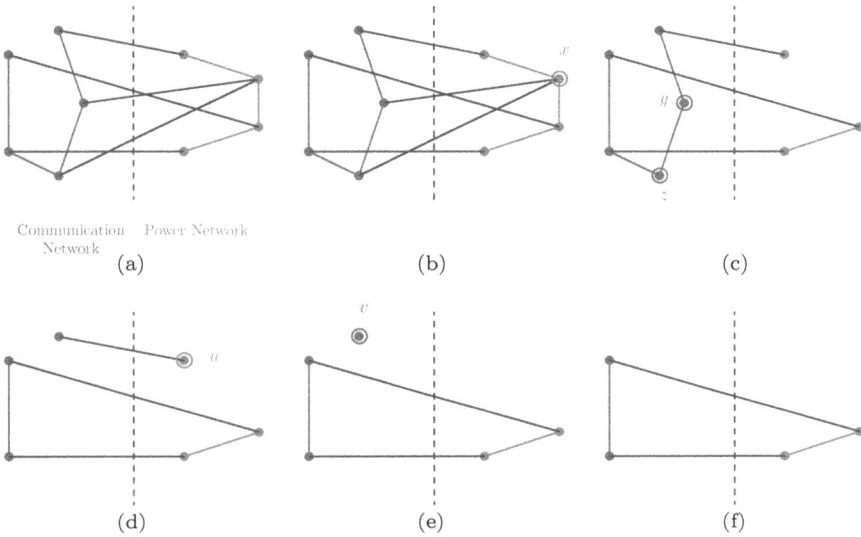

Figure 2. Cascading failures in interdependent smart grid networks. The faulty nodes are shown as double circles.

6.1. *Stage I: Targeted attack on the communication network*

We consider three types of targeted attacks. A node is removed with probability proportional to its degree, betweenness, or clustering coefficient. We assume that the attacker uses only one of these centrality measures for node compromise, but not a combination of all three. Let ϕ_k be the probability that a node i of centrality measure k (either degree or betweenness or clustering coefficient of node i) is not removed. Clearly,

$$\phi_k = 1 - \frac{\text{Centrality}(i)}{\sum_{v \in V_A} \text{Centrality}(v)}.$$

For example, $\deg(i) = Ak^{-\alpha_A}$ is the degree of node i and m_A is the number of edges in N_A. If the attacker decides to remove nodes with high degree then,

$$\phi_k = 1 - \frac{Ak^{-\alpha_A}}{2m_A}.$$

We note that $\alpha_A = 0$ represents random removal of nodes.

We will first calculate the size of the giant component $\mathcal{G}A_1$. Let u denote the average probability that a node is not connected to the giant cluster via one of its neighbors. Consider a node of degree k. The probability that it is not connected to the giant component via any of its neighbors is u^k.

Probability of it being in the giant component
    = Probability that it is not attacked
       · Probability that one of its neighbors is in the giant component.

Thus, the probability of it being in the giant component is $\phi_k(1 - u^k)$. Averaging over the degree distribution $P_A(k)$, we can calculate the fraction of nodes in the initial giant component as

$$\mu_{A_1} = \sum_{k=0}^{\infty} P_A(k)\phi_k(1 - u^k)$$

$$= \sum_{k=0}^{\infty} P_A(k)\phi_k - \sum_{k=0}^{\infty} P_A(k)\phi_k u^k \tag{6}$$

$$= f_0(1) - f_0(u),$$

where

$$f_0(z) = \sum_{k=0}^{\infty} P_A(k)\phi_k z^k. \tag{7}$$

We will now show how to calculate u. A node is not connected to the giant component when either of the following cases arises:

- The node is attacked and thus removed.
- The node is present, but not connected to any node in the giant component.

Let k be the excess degree of a neighboring node. The original degree of a node is one more than the excess degree, i.e., $k + 1$ [37]. The probability that a neighbor is removed is $1 - \phi_{k+1}$. The probability that a neighbor is present, but the node itself is not present in the giant component is $\phi_{k+1}u^k$.

Hence, using [37], u can be calculated as,

$$u = \sum_{k=0}^{\infty} q_{A_k}(1 - \phi_{k+1} + \phi_{k+1}u^k)$$
$$= 1 - f_1(1) + f_1(u), \tag{8}$$

where

$$f_1(z) = \sum_{k=0}^{\infty} q_{A_k}\phi_{k+1}z^k. \tag{9}$$

Note that q_{Ak}, the probability of a node having excess degree k in N_A, can be given by $q_{Ak} = \frac{(k+1)P_A(k+1)}{\langle k_A \rangle}$ [37]. It can be seen that $\sum_{k=0}^{\infty} q_A = 1$. Substituting the value of q_{Ak+1}, the value of $f_1(z)$ can be calculated as,

$$f_1(z) = \sum_{k=0}^{\infty} \frac{(k+1)P_A(k+1)}{\langle k_A \rangle}\phi_{k+1}z^k$$
$$= \frac{1}{\langle k_A \rangle} \sum_{k=0}^{\infty} kP_A(k)\phi_k z^{k-1}, \tag{10}$$

where $\langle k_A \rangle$ is the average degree of nodes in N_A. We observe that

$$f_1(z) = \frac{f'0(z)}{g'_{A_0}(1)}, \tag{11}$$

where $g_{A_0}(z)$ is the generating function

$$g_{A_0}(z) = \sum_{k=0}^{\infty} P_A(k)z^k. \tag{12}$$

6.2. Stage II: Effect of cascading failure on the power network

In a power grid, the effect of a communication network is not so pronounced. However, in a smart grid, the communication and power networks reinforce each other. Due to attack on nodes in the communication network, some nodes in the power network might be affected. This happens for smart grids, which are interconnected networks.

A node in the power network N_B is functional if a node and in N_B has at least one support link from N_A. Initially, at stage II, all nodes in N_B are in the giant component. We consider all those nodes which are supported by nodes not in $\mathcal{G}A_1$. Such nodes will not remain functional because they will be cut off from the communication network. The probability that a node is not in the giant component $\mathcal{G}A_1$ is $1 - \mu_A$. Suppose, a node is supported by $\tilde{k}B$ nodes in N_A. The probability that the k_B neighboring nodes are not in $\mathcal{G}A_1$ is $(1 - \mu_A)^{\tilde{k}_B}$. The fraction of nodes in N_B disconnected due to attack on N_A is given by

$$r_{B_2} = \sum_{\tilde{k}B=0}^{\infty} \tilde{P}_B(\tilde{k}B)(1 - \mu_{A_1})^{\tilde{k}_B}. \tag{13}$$

The fraction of nodes remaining in N_B is given by $1 - r_{B_2}$. This is similar to the random removal of vertices. The fraction of nodes in the resulting giant component can be calculated by the technique in

$$\mu_{B_2} = (1 - r_{B_2})(1 - g_{B_0}(u)), \tag{14}$$

where

$$u = 1 - \phi + \phi g_{B_1}(u), \tag{15}$$

$$g_{B_c}(u) = \sum_{k=0}^{\infty} P_B(k)u^k \tag{16}$$

and

$$g_{B_1}(z) = \sum_{k=0}^{\infty} q_{B_k} z^k. \tag{17}$$

6.3. *Stage III: Cascading failure in the communication network*

We will now study the effect of cascading failure in the communication network due to the failure in power networks. Each node in

N_A is supported by only one link from the power network. If a node in N_B fails, then the communication node it supports also fails. We have assumed a simple interconnection pattern for ease of analysis. For more complex interconnection patterns, only the links connecting the failed node in N_B are disrupted. The fraction of nodes in N_A which fail due to failure of the node in N_B is given by

$$r_{A3} = \sum_{\tilde{k}_A=0}^{\infty} \tilde{P}_A(\tilde{k}_A)(1 - \mu_{B_2}). \tag{18}$$

We can consider that these nodes are randomly removed in N_A and find the giant component resulting due to this removal of nodes. The fraction of nodes in the giant component which result from this random compromise is calculated as shown in Section 5 as

$$\mu_{A3} = (1 - r_{A3})(1 - g_{A_0}(u)), \tag{19}$$

where

$$u = 1 - r_{A3} + r_{A3}g_{A_1}(u), \tag{20}$$

$$g_{A_0}(u) = \sum_{k=0}^{\infty} P_A(k)u^k \tag{21}$$

and

$$g_{A_1}(z) = \sum_{k=0}^{\infty} qA_k z^k. \tag{22}$$

6.4. *Stage IV: Cascading failure in the power network*

We now calculate the number of nodes in the power network which are connected to nodes not in the giant component in the communication network. The fraction of nodes which are removed because they have all their support links from the nodes not in the giant component of N_A is given by

$$r_{B_4} = \sum_{k_{\tilde{B}}=0}^{\infty} \tilde{P}_B(\tilde{k}_B)(1 - \mu_{A_3})^{\tilde{k}_B}. \tag{23}$$

The giant component can be calculated as in Section 5.

6.5. *Giant components and steady-state conditions*

We will now calculate the size of the giant component at steady state. Let, $r_{A_{2n-1}}(n \geq 1)$ be the fraction of nodes in N_A that are removed due to the removal of nodes in N_B at stage $2n - 2$. For $n = 1$, the analysis is given in Section 6.1. Then,

$$r_{A_{2n-1}} = \sum_{\tilde{k}_A=0}^{\infty} (1 - \mu_{B_{2n-2}})\tilde{P}_A(\tilde{k}_A). \tag{24}$$

Proceeding in the same way as that mentioned earlier, the general expression for nodes for the fraction of nodes in the giant component at the $(2n - 1)$th stage in the communication network is given by

$$\mu_{A_{2n-1}} = (1 - r_{A_{2n-1}})(1 - g_{A_0}(u)), \tag{25}$$

where

$$u = 1 - \phi_{A_{2n-1}} + \phi_{A_{2n-1}}g_{A_1}(u), \tag{26}$$

$$g_{A_0}(u) = \sum_{k=0}^{\infty} P_A(k)u^k, \tag{27}$$

and

$$g_{A_1}(z) = \sum_{k=0}^{\infty} q_{A_k}z^k. \tag{28}$$

Similarly, let, $r_{B_{2n}}$ be the fraction of nodes in N_B that are removed due to the removal of nodes in N_A at stage $2n - 1$. Then,

$$r_{B_{2n}} = \sum_{\tilde{k}_B=0}^{\infty} \tilde{P}_B(\tilde{k}_B)(1 - \mu_{A_{2n-1}})\tilde{k}_B \tag{29}$$

The fraction of nodes in the giant component of N_B at stage $2n$ is given by

$$\mu_{B_{2n}} = (1 - r_{B_{2n}})(1 - g_{B_0}(u)), \tag{30}$$

where

$$u = 1 - \phi + \phi g_{A_1}(u), \tag{31}$$

$$g_{B_0}(u) = \sum_{k=0}^{\infty} P_B(k)u^k, \tag{32}$$

and

$$g_{B_1}(z) = \sum_{k=0}^{\infty} q_{B_k} z^k. \tag{33}$$

We arrive at a steady state when

$$\mu_{A_{2n-1}} = \mu_{A_{2n+1}} = \mu_{A_{2n+3}} = \cdots \tag{34}$$

$$\mu_{B_{2n-2}} = \mu_{B_{2n}} = \mu_{B_{2n+2}} = \cdots \tag{35}$$

It is difficult to solve these systems of equations analytically. So, we generate the smart grid using different random graph models and simulate the effect of targeted and random attacks on these graphs. The results of this study are given in the next section.

6.6. *Random attacks*

In random attacks, the attacker chooses the nodes of a network either uniformly at random or according to a probability distribution defined on the nodes. If the network has n nodes, the attacker chooses each node of the communication network with a probability of $\frac{1}{n}$ (for the uniformly-at-random case). This causes a cascading failure in the power network and the process is repeated.

6.7. *Adaptive attacks*

In adaptive attacks, the attacker deletes nodes iteratively in rounds, instead of all nodes at once. In each round, the attacker chooses a set of nodes to be compromised based on the new centrality measure and removes this set. Since the adversary makes his decision based on the result in the previous round, it is evident that he can disrupt the connectivity of the network more effectively. It has been observed through experiments that an adversary has an advantage while compromising nodes adaptively, compared to non-adaptive deletion. The paper [48] discusses some strategies for defending against adaptive attacks.

7.   Experimental Results

7.1.   *Experimental set-up*

In order to simulate a smart grid, we use the network library *igraph* [12] on C. Since previous studies [39] have shown that power grids follow a power law degree distribution, we have considered the power network as an SF network and compared it with ER networks. We consider two networks, the power network and the communication network, both of which are SF networks. For each communication node, an interlink is assigned by choosing a power node at random. We consider three types of attack on the communication network — targeted, random, and mixed (combination of the first two). In the random attack, we choose x nodes uniformly at random from all the nodes without replacement. In the targeted attack, we choose x nodes without replacement, such that the probability of choosing a node is proportional to the degree. For mixed attacks, we select half of the nodes for targeted attack and half of the nodes for random attack. We evaluate the resilience of networks based on the size of the giant components and average path length under attack, in either of the networks (power and communication). Path length is an important parameter, because the longer the path length, the larger the communication overheads and delay in message transmission. We study targeted attacks by compromising nodes based on degree, betweenness, and clustering coefficients. Finally, we study the effect of compromise by running the experiment 50 times for each input x. The same graphs are considered every time.

We compare our results with interdependent networks, with the ER–ER combination instead of the SF–SF combination. We also study adaptive attacks.

7.2. *Experimental results, observations, and inferences*

In Figures 3–13, we present our experimental results. All experiments reported in Figures 4–9 are performed on communication networks with 2,000 nodes and power networks with 1,000 nodes.

In Figure 3, the power network consists of 1,000 nodes and the communication network consists of 10,000 nodes. The communication/power network is generated as an SF network using a power

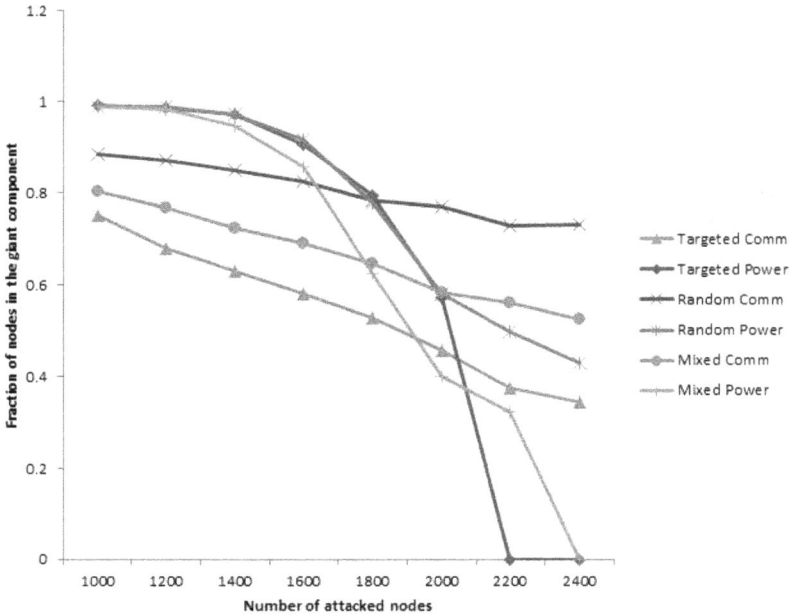

Figure 3. Variation of giant component size with number of attacked nodes for targeted, random, and mixed attacks.

law degree distribution. We have plotted the size of the giant component (as a fraction of the size of the communication/power network) against the number of nodes attacked in the communication network. We observe that for a given value of the number of attacked nodes (only nodes in the communication network are attacked), the fraction of nodes in the giant component of the communication network is highest for random attacks and lowest for targeted attack (based on degree). The corresponding fraction for mixed attacks lies somewhere in the middle. We also see that for the same fraction of nodes compromised, the giant component of the power network disintegrates faster for targeted attacks compared to random attacks. We see that on compromising 2,200 nodes, there is no giant component when targeted attack occurs, whereas a giant component exists under random attacks. This is expected, as attacking higher-degree nodes results in a faster disintegration of the network, resulting in smaller components.

In Figures 4 and 5, the communication network consists of 2,000 nodes, whereas the power network consists of 1,000 nodes. We compare the results for a combination of SF–SF and ER–ER networks. In Figure 4, we have plotted the size of the giant component (as a fraction of the size of the communication network) against the number of attacked nodes. The targeted attack is based on the degree of nodes, i.e., high-degree nodes are compromised with high probability. The communication network is generated using (i) an SF network using a power law degree distribution, (ii) the ER $G(n,p)$ model with $p = 0.01$, and (iii) the ER $G(n,p)$ model with $p = 0.005$. Only nodes in the communication network are attacked. In Figure 5, we have plotted the size of the giant component (as a fraction of the size of the power network) against the number of attacked nodes (based on degree). The power network is generated using the same models as mentioned earlier. From Figures 4 and 5, we see that, for all three types of attacks, the sizes of the giant components for ER graphs are comparable. The power and communication networks will be more fault tolerant to targeted attacks for ER networks, compared to SF networks.

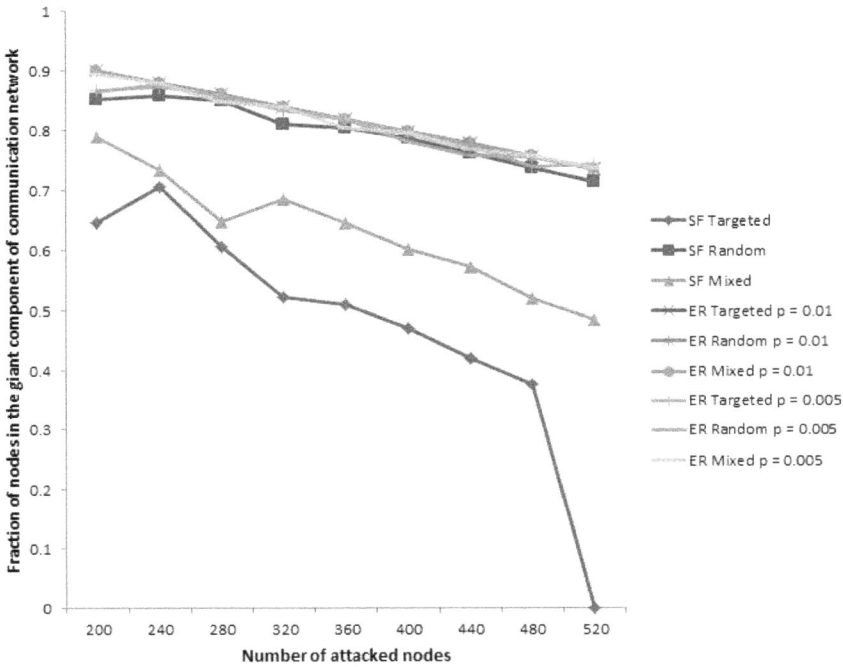

Figure 4. Variation of giant component size with number of attacked nodes in the communication network for scale-free and Erdos–Renyi models.

In Figures 6 and 7, the communication network again consists of 2,000 nodes, whereas the power network consists of 1,000 nodes. In Figure 6, we have plotted the size of the giant component (as a fraction of the size of the communication network) against the number of attacked nodes for various types of attacks. We have considered two different types of networks SF and ER with $p = 0.01$. For the same SF networks, we have considered attack of nodes based on their (i) degree, (ii) betweenness, and (iii) clustering coefficients, and compared them with random attacks. We have done a similar study for ER networks and considered the different types of attacks as above for the same network. We found that, to cause the maximum damage (get the smallest giant component) to an SF network, an adversary must delete high betweenness nodes with high probability. The next best strategy for an attacker is to compromise high-degree

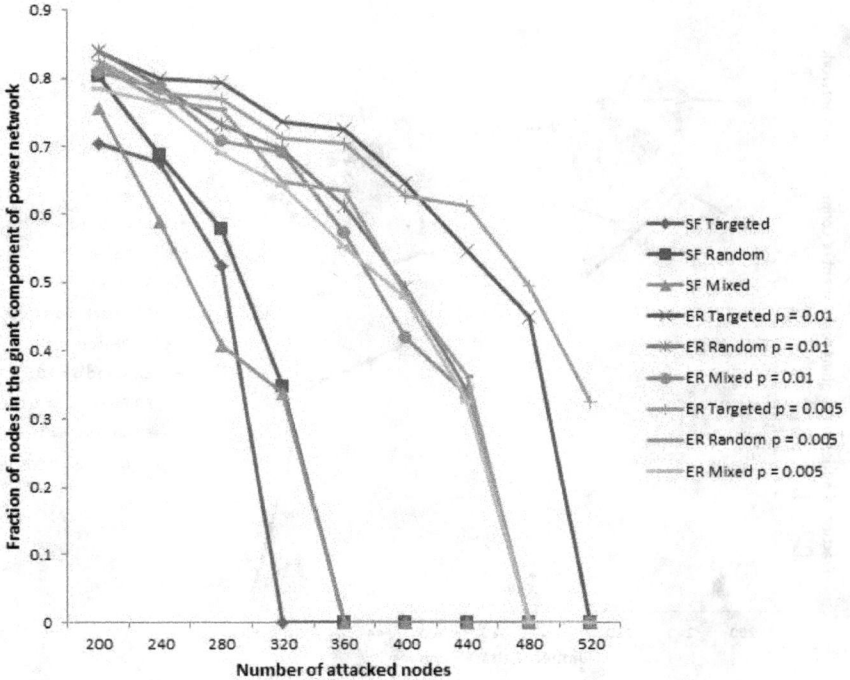

Figure 5. Variation of giant component size with number of attacked nodes in the power network for scale-free and Erdos–Renyi models.

nodes with high probability. The third best strategy is to compromise nodes with high clustering coefficient with high probability. For ER networks, the attacker can use any attack strategy, because all of them give approximately the same results. In Figure 7, we have plotted the size of the giant component (as a fraction of the size of the power network) against the number of attacked nodes for various types of attacks (based on degree, betweenness, and clustering coefficients). Here, we can see that for both SF and ER networks, targeted attacks based on betweenness are the most effective strategy, while random attacks are the least effective from the point of view of an adversary. We observe that after targeted attacks, the size of the giant component in ER networks can be twice as large as that in SF networks.

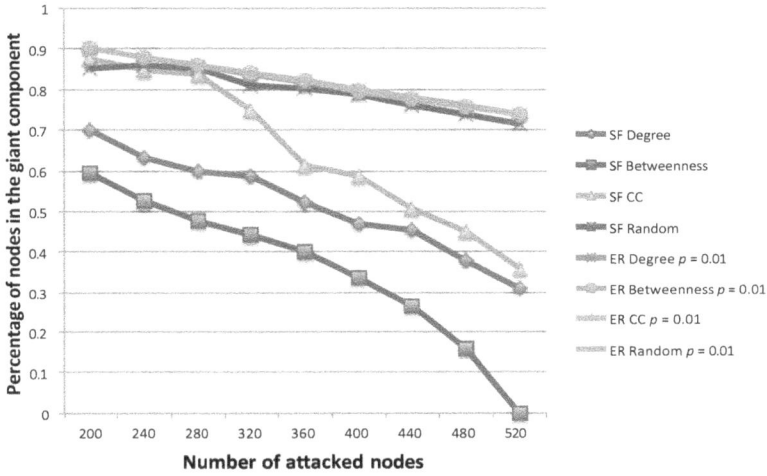

Figure 6. Variation of giant component size of communication network with number of attacked nodes under different attack models, in the communication network for scale-free and Erdos–Renyi models.

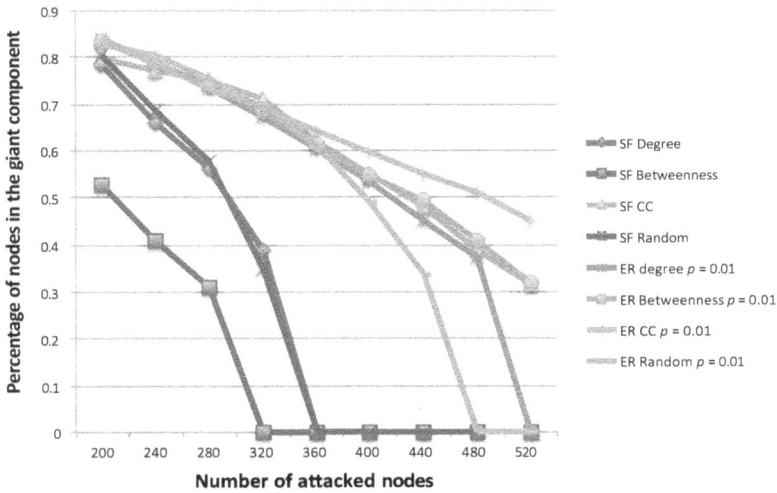

Figure 7. Variation of giant component size of power network with number of attacked nodes in the power network for scale-free and Erdos–Renyi models.

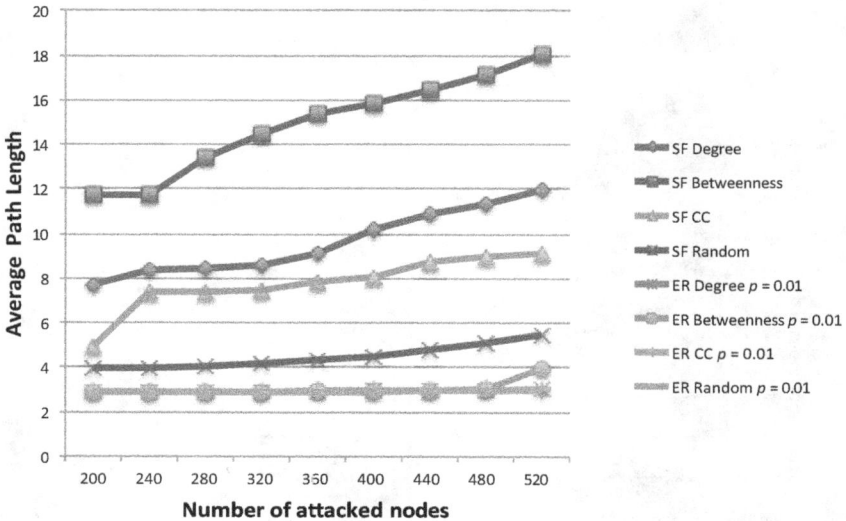

Figure 8. Variation of average path length in communication network with number of attacked nodes in the network for scale-free and Erdos–Renyi models.

In Figures 8 and 9, the communication network again consists of 2,000 nodes, whereas the power network consists of 1,000 nodes. The average path length is a measure of how many hops a message must travel to reach the destination and should be minimized, in order to reduce the communication overhead. In Figure 8, we compare the average path length due to node compromise for the communication network. We have considered only the average path length in the giant component, because that is the largest functional component in the network. For the same SF networks, we have considered following attack of nodes based on their (i) degree, (ii) betweenness, and (iii) clustering coefficients, and comparing them with random attacks. We have done a similar study for ER network and considered the different types of attacks as above for the same network. For the SF network, we see that on compromising nodes with probability proportional to the betweenness, the average path length increases the most. The average path length is the longest for case (ii), followed by (i) and (iii). For ER networks, the path length for all three cases remains almost the same. In Figure 9, we compare the average path

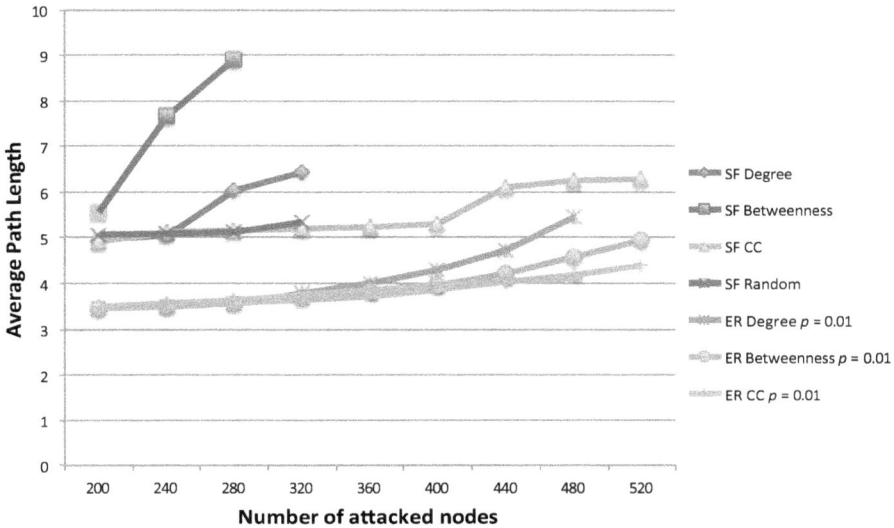

Figure 9. Variation of average path length in power network with number of attacked nodes in the network for scale-free and Erdos–Renyi models.

length due to node compromise for the power network. In this case, the average path length is highest for SF networks when the attack is done based on betweenness and lowest for attack based on clustering coefficient. For ER graphs, the average path length is highest for attack based on degree and lowest for attack based on betweenness. In Figure 9, the average path length has not been plotted for node compromise beyond 320 nodes in SF–SF networks (for compromise based on degree/betweenness/clustering coefficients) because the giant component vanishes at this point.

We have carried out experiments with real power grid data for Western States power grid [52]. We have coupled this network with the synthetic SF communication network. The Western States power grid consists of 4,941 nodes, whereas the communication network consists of 9,000 nodes. We compare a random attack with a targeted attack based on degree, betweenness, and clustering coefficient. As in the previous case, the best attack strategy for the attacker is to compromise nodes based on betweenness and the worst strategy is

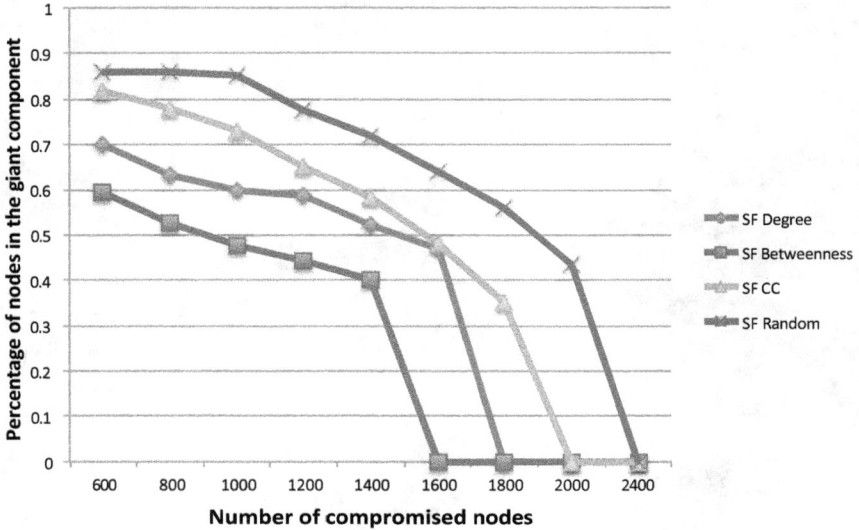

Figure 10. Variation of giant component size with number of attacked nodes in the communication network for Western States power grid coupled with simulated communication network.

to compromise nodes randomly. The results are shown in Figures 10 and 11.

In Figure 12, we present experimental results for adaptive attacks. We consider communication and power SF networks for sizes 10,000 and 1,000 each. We compromise nodes in the following ways:

(1) We consider adaptive attacks, in which we compromise 200 nodes at a time, every time calculating the new set of compromised nodes, with probability proportional to the degree at that instant.

(2) We consider attacks where 1,000 nodes are compromised together, with probability proportional to the degree.

In Figure 13, we compromise nodes selectively and adaptively based on the betweenness. Similar to that mentioned earlier, the adaptive node compromise results in a smaller giant component than non-adaptive node compromise. Also, compromising nodes selectively based on betweenness of nodes result in a smaller size of the

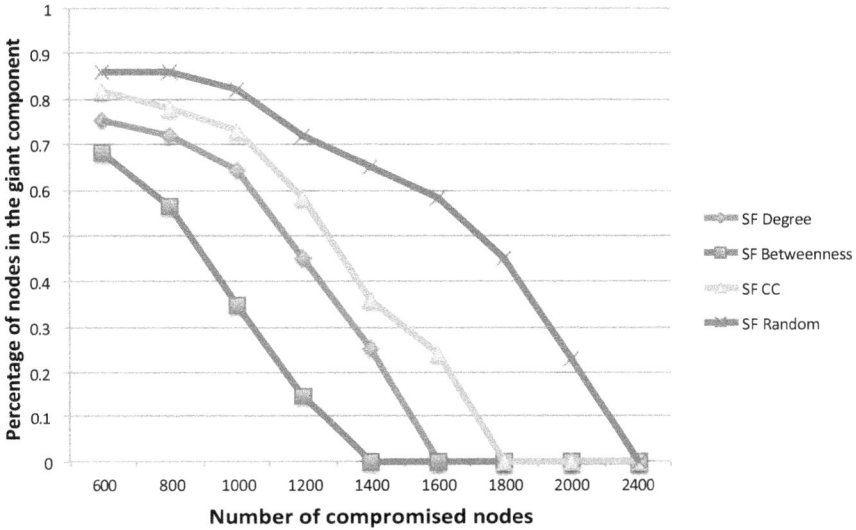

Figure 11. Variation of giant component size with number of attacked nodes in the power network for Western States power grid coupled with simulated communication network.

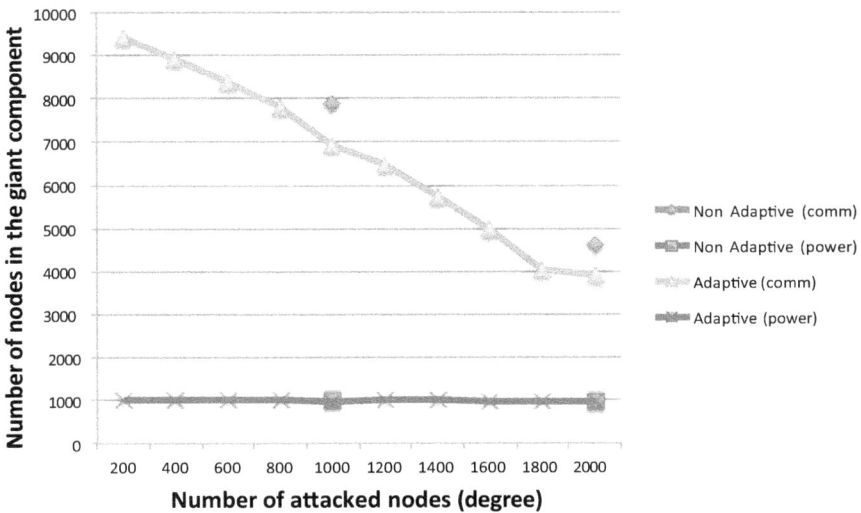

Figure 12. Variation of giant component size of communication and power network with number of attacked nodes (based on degree) for adaptive and non-adaptive deletion.

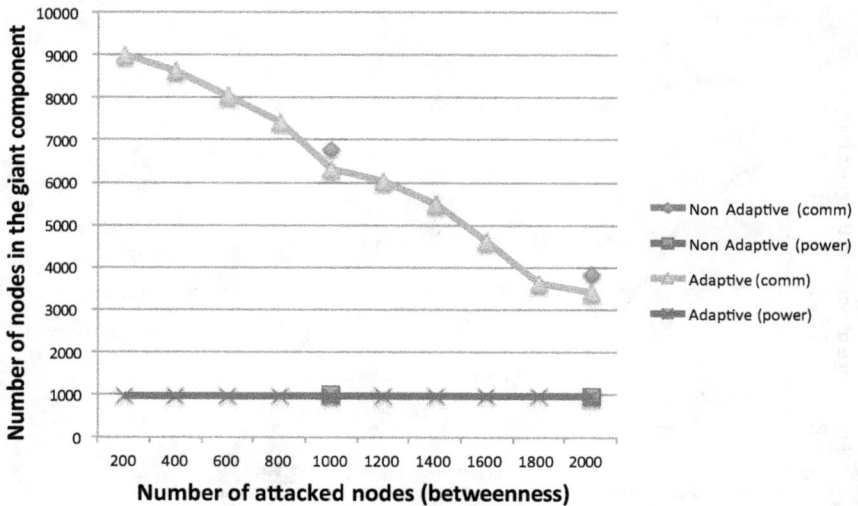

Figure 13. Variation of giant component size of communication and power network with number of attacked nodes (based on betweenness) for adaptive and non-adaptive deletion.

giant component compared to that based on degree (as we have seen before). For example, when 1,000 nodes are compromised adaptively based on degree, the size of the giant component in communication network is 6,915, whereas compromising adaptively based on betweenness results in a giant component of 6,310 nodes in the communication network. When nodes are compromised adaptively based on clustering coefficient, the adversary has less advantage compared to compromising adaptively based on the degree.

7.3. *Strategies for defending smart grids*

Although we have mostly talked about how an attacker can target vulnerable nodes in a network, we can use this to design a robust and fault-tolerant network. Understanding attack strategies is important in designing good defense strategies. Since it is advantageous for an attacker to do targeted attacks on nodes with high betweenness centrality, guarding these nodes (power stations and communication centers) is vital for the reliability of a smart grid. ER networks are more robust to attacks than SF networks. Even with the current security measures taken by governments across the world, cyber-security of critical infrastructure is not fool-proof. This chapter stresses the importance of protecting the important nodes in the network because of their vulnerabilities to attack. Though most of the power grid networks have a power law degree distribution [39], it might be better to design the underlying network as an ER network. This will result in more robust networks, though at the cost of efficiency.

8. Conclusion and Future Work

In this chapter, we have modeled the power and communication networks in a smart grid as two interdependent networks, and analyzed the cascading failure in smart grids for targeted attacks. This is one of the early works on targeted attacks and adaptive attacks on smart grids, which are modeled as interdependent networks. We have given a mathematical expression for the size of the giant component when nodes are compromised. We have carried out extensive experiments to show that targeted attacks give an advantage to the adversary over random attacks and adaptive attacks are more effective than non-adaptive attacks. For targeted attacks, choosing a node with a probability proportional to its betweenness is more effective than doing so with degree or clustering coefficient. We have shown that SF–SF networks are more prone to targeted attacks than ER–ER networks.

A challenging open problem is to obtain a closed-form solution for the size of the giant component from the mathematical analysis that we have presented. Another important question is to present a good model of smart grids, which will be resilient to both random and targeted attacks. The structure of both the power and communication networks and the assignment of interlinks need to be studied. An important question is to find which network model and interconnection pattern will increase the resilience of the smart grid. Since the smart grid will be a component of the Internet of Things (IoT), a future direction of work is to propose a model which will be re-silient to attacks and can disseminate information rapidly in the network.

References

[1] Ács, G. and Castelluccia, C. I have a dream! (differentially private smart metering). In *Information Hiding — 13th International Conference, IH 2011*, Prague, Czech Republic, May 18–20, 2011, Revised Selected Papers, pp. 118–132, 2011.

[2] Bader, D.A. and Madduri, K. Parallel algorithms for evaluating centrality indices in real-world networks. In *2006 International Conference on Parallel Processing (ICPP 2006)*, August 14–18, 2006, Columbus, Ohio, USA, pp. 539–550. IEEE Computer Society, 2006.

[3] Bobba, R., Khurana, H., AlTurki, M., and Ashraf, F. PBES: A policy based encryption system with application to data sharing in the power grid. In *ASIACCS*, pp. 262–275, 2009.

[4] Bose, A. Smart transmission grid applications and their supporting infrastructure. *IEEE Transactions on Smart Grids*, 1(1):11–19, 2010.

[5] Brandes, U. A faster algorithm for betweenness centrality. *Journal of Mathematical Sociology*, 25(2):163–177, 2001.

[6] Brandes, U. and Pich, C. Centrality estimation in large networks. *International Journal of Bifurcation and Chaos*, 17(1):2303–2318, 2007.

[7] Yağan, O. and Gligor, V. Analysis of complex contagions in random multiplex networks. *Physical Review E*, 86:036103, 2012.

[8] Buldyrev, S.V., Parshani, R., Paul, G., Eugene Stanley, H., and Havlin, S. Catastrophic cascade of failures in interdependent networks. *Nature*, 464:1025–1028, 2010.

[9] Buldyrev, S.V., Shere, N.W., and Cwilich, G.A. Interdependent networks with identical degrees of mutually dependent nodes. *Physical Review E*, 83:016112, 2011.

[10] Chen, X., Dinh, H., and Wang, B. Cascading failures in smart grid - benefits of distributed generation. In *SmartGridComm*, pp. 73–78, 2010.

[11] Chertkov, M., Pan, F., and Stepanov, M.G. Predicting failures in power grids: The case of static overloads. *IEEE Transactions on Smart Grid*, 2(1):162–172, 2011.

[12] Csardi, G. and Nepusz, T. The igraph software package for complex network research (http://igraph.sourceforge.net/). *InterJournal Complex Systems*, 1695, 2006.

[13] Dong, G., Gao, J., Du, R., Tian, L., Eugene Stanley, H., and Havlin, S. Robustness of network of networks under targeted attack. *Physical Review E*, 87:052804, 2013.

[14] Durrett, R. *Random Graph Dynamics*. Cambridge University Press, Cambridge, 2007.

[15] Fouda, M., Fadlullah, Z.M., Kato, N., Lu, R., and Shen, X. A lightweight message authentication scheme for smart grid communications. *IEEE Transactions on Smart Grid*, 2(4):675–685, 2011.

[16] Gupta, S.R., Kazi, F.S., Wagh, S.R., and Singh, N.M. Probabilistic framework for evaluation of smart grid resilience of cascade failure. In *2014 IEEE Innovative Smart Grid Technologies-Asia (ISGT ASIA)*, pp. 255–260. IEEE, 2014.

[17] Huang, X., Gao, J., Buldyrev, S.V., Havlin, S., and Eugene Stanley, H. Robustness of interdependent networks under targeted attack. *Physical Review E*, 83:065101, 2011.

[18] Huang, Z., Wang, C., Ruj, S., Stojmenovic, M., and Nayak, A. Modeling cascading failures in smart power grid using interdependent complex networks and percolation theory. In *The 8th IEEE Conference on Industrial Electronics and Applications, ICIEA'13*, pp. 1023–1028, 2013.

[19] Huang, Z., Wang, C., Stojmenovic, M., and Nayak, A. Percolation of partially interdependent networks under targeted attack. *IEEE Transactions on Emerging Topics in Computing*, 64(8), 2158–2168. 2013.

[20] Huang, Z., Wang, C., Stojmenovic, M., and Nayak, A. Characterization of cascading failures in interdependent cyber-physical systems. *Transactions on Computers*, (99), 2015.

[21] Jiang, H., Wang, Z., and He, H. An evolutionary computation approach for smart grid cascading failure vulnerability analysis. In *2019 IEEE Symposium Series on Computational Intelligence (SSCI)*, pp. 332–338. IEEE, 2019.

[22] Kadloor, S. and Santhi, N. Understanding cascading failures in power grids. *CoRR*, abs/1011.4098, 2010.

[23] Kang, U., Papadimitriou, S., Sun, J., and Tong, H. Centralities in large networks: Algorithms and observations. In *Proceedings of the Eleventh SIAM International Conference on Data Mining, SDM 2011*, April 28–30, 2011, Mesa, Arizona, USA, pp. 119–130. SIAM/Omnipress, 2011.

[24] Kas, M. *Incremental Centrality Algorithms for Dynamic Network Analysis*. Ph.D. Thesis, Carnegie Mellon University, 2013.

[25] Kgwadi, M. and Kunz, T. Securing RDS broadcast messages for smart grid applications. In Helmy, A., Mueller, P., and Zhang, Y. (Eds.), *IWCMC*, pp. 1177–1181. ACM, 2010.

[26] Kushner, D. The real story of stuxnet. *IEEE Spectrum*, March 2013.

[27] Lee, A. and Brewer, T. Smart grid cyber security strategy and requirements. *NISTIR 7628*, February 2009.

[28] Li, F., Luo, B., and Liu, P. Secure information aggregation for smart grids using homomorphic encryption. In *2010 first IEEE International Conference on Smart Grid Communications*, pp. 327–332. IEEE SmartGridComm, 2010.

[29] Liu, J., Xiao, Y., Li, S., Liang, W., and Philip Chen, C.L. Cyber security and privacy issues in smart grids. *IEEE Communications Surveys and Tutorials*, 14(4):981–997, 2012.

[30] Liu, S., Chen, B., Zourntos, T., Kundur, D., and Butler-Purry, K.L. A coordinated multi-switch attack for cascading failures in smart grid. *IEEE Transactions on Smart Grid*, 5(3):1183–1195, 2014.

[31] Liu, S., Kundur, D., Zourntos, T., and Butler-Purry, K.L. Coordinated variable structure switching attack in the presence of model error and state estimation. In *IEEE Third International Conference on Smart Grid Communications, SmartGridComm 2012*, Tainan, Taiwan, November 5–8, 2012, pp. 318–323. IEEE, 2012.

[32] Liu, S., Mashayekh, S., Kundur, D., Zourntos, T., and Butler-Purry, K.L. A framework for modeling cyber-physical switching attacks in smart grid. *IEEE Transactions on Emerging Topics Computing*, 1(2):273–285, 2013.

[33] Lu, R., Liang, X., Li, X., Lin, X., and Shen, X. EPPA: An efficient and privacy-preserving aggregation scheme for secure smart grid communications. *IEEE Transactions on Parallel and Distributed Systems*, 23(9):1621–1631, 2012.

[34] Lu, R., Lin, X., Shi, Z., and Shen, X. EATH: An efficient aggregate authentication protocol for smart grid communications. In *2013 IEEE Wireless Communications and Networking Conference (WCNC)*, Shanghai, China, April 7–10, 2013, pp. 1819–1824. IEEE, 2013.

[35] McLaughlin, A. and Bader, D.A. Scalable and high performance betweenness centrality on the GPU. In *International Conference for High Performance Computing, Networking, Storage and Analysis, SC 2014*, New Orleans, LA, USA, November 16–21, 2014, pp. 572–583. IEEE, 2014.

[36] Molina-Markham, A., Danezis, G., Fu, K., Shenoy, P.J., and Irwin, D.E. Designing privacy-preserving smart meters with low-cost microcontrollers. In Keromytis, A.D. (Ed.), *Financial Cryptography and Data Security - 16th International Conference, FC 2012*, Kralendijk, Bonaire, February 27–March 2, 2012, Revised Selected Papers, volume 7397 of *Lecture Notes in Computer Science*, pp. 239–253. Springer, 2012.

[37] Newman, M.E.J. *Networks: An Introduction.* Oxford University Press, Oxford, 2010.

[38] Nguyen, D.T., Shen, Y., and Thai, M.T. Detecting critical nodes in interdependent power networks for vulnerability assessment. *IEEE Transactions on Smart Grid*, 4(1):151–159, 2013.

[39] Pagani, G.A. and Aiello, M. The power grid as a complex network: A survey. *Physica A: Statistical Mechanics and Its Applications*, 392:2688–2700, 2013.

[40] Pfitzner, R., Turitsyn, K.S., and Chertkov, M. Controlled tripping of overheated lines mitigates power outages. *CoRR*, abs/1104.4558, 2011.

[41] Rad, A.H.M., Wong, V.W.S., Jatskevich, J., Schober, R., and Leon-Garcia, A. Autonomous demand-side management based on game-theoretic energy consumption scheduling for the future smart grid. *IEEE Transactions on Smart Grid*, 1(3):320–331, 2010.

[42] Rial, A. and Danezis, G. Privacy-preserving smart metering, 2010. Technical Report MSR-TR-2010-150, Microsoft Research.

[43] Ruj, S. and Nayak, A. A decentralized security framework for data aggregation and access control in smart grids. *IEEE Transactions on Smart Grid*, 4(1):196–205, 2013.

[44] Ruj, S. and Pal, A. Analyzing cascading failures in smart grids under random and targeted attacks. In *28th IEEE International Conference on Advanced Information Networking and Applications, AINA 2014*, Victoria, BC, Canada, May 13–16, 2014, pp. 226–233. IEEE Computer Society, 2014.

[45] Ruj, S. and Pal, A. *Fault Tolerance and Reliability of Smart Grids*, Encyclopedia of Wireless Networks, pp. 1–11. Springer International Publishing, Cham, 2019.

[46] Sen, A., Mazumder, A., Banerjee, J., Das, A., and Compton, R. Identification of K most vulnerable nodes in multi-layered network using a new model of interdependency. In *2014 Proceedings IEEE INFOCOM Workshops*, Toronto, ON, Canada, April 27–May 2, 2014, pp. 831–836. IEEE, 2014.

[47] Shao, J., Buldyrev, S.V., Havlin, S., and Eugene Stanley, H. Cascade of failures in coupled network systems with multiple support-dependent relations. *CoRR*, 2010.

[48] Wang, W., Ma, Q., Lin, L., Guan, T., Tian, D., Li, S., and Li, J. Adaptive observer-based attack location and defense strategy in smart grid. *Proceedings of the Institution of Mechanical Engineers, Part I: Journal of Systems and Control Engineering*, 236(2):294–304, 2022.

[49] Wang, W. and Lu, Z. Cyber security in the smart grid: Survey and challenges. *Computer Networks*, 57(5):1344–1371, 2013.

[50] Wang, W., Xu, Y., and Khanna, M. A survey on the communication architectures in smart grid. *Computer Networks*, 55(15):3604–3629, 2011.

[51] Wang, W., Xu, Y., and Khanna, M. Enhancing network robustness against malicious attacks. *Physical Review E*, 85:066130, 2012.

[52] Watts, D.J. and Strogatz, S.H. Collective dynamics of 'small-world' networks. *Nature*, 393:440–442, 1998.

[53] Yagan, O., Qian, D., Zhang, J., and Cochran, D. Optimal allocation of interconnecting links in cyber-physical systems: Interdependence, cascading failures, and robustness. *IEEE Transactions on Parallel and Distributed Systems*, 23(9):1708–1720, 2012.

[54] Yuan, Y., Li, Z., and Ren, K. Modeling load redistribution attacks in power systems. *IEEE Transactions on Smart Grid*, 2(2):382–390, 2011.

[55] Yuan, Y., Li, Z., and Ren, K. Quantitative analysis of load redistribution attacks in power systems. *IEEE Transactions on Parallel and Distributed Systems*, 23(9):1731–1738, 2012.

[56] Zhu, W., Han, M., Milanović, J.V., and Crossley, P. Methodology for reliability assessment of smart grid considering risk of failure of communication architecture. *IEEE Transactions on Smart Grid*, 11(5):4358–4365, 2020.

[57] Zúñiga, A.A., Baleia, A., Fernandes, J., and Branco, P.J.D.C. Classical failure modes and effects analysis in the context of smart grid cyber-physical systems. *Energies*, 13(5):1215, 2020.

https://doi.org/10.1142/9789811273551_0007

Chapter 7

Curricular Guidance to Bridge the IT–OT Cybersecurity Gap

Sean McBride[*,‡], **Corey Schou**[*,§], **and Jill Slay**[†,¶]

Idaho State University
†*University of South Australia*
‡*seanmcbride@isu.edu*
§*schou@iri.isu.edu*
¶*jill.slay@unisa.edu.au*

Abstract: Professionals and academics alike are becoming cognizant about the security ramifications of key differences between information technology (IT) and operational technology (OT). This paper presents work to create a knowledge unit to guide the development of cybersecurity professionals who comfortably interact with both IT and OT systems.

Keywords: Industrial Cybersecurity, Education and Training, IT–OT Gap.

1. Introduction

Professionals and academics feel comfortable with the ubiquitous information technology (IT) intended to make their lives more productive and enjoyable. Email, apps, video calls, servers, memory, and bandwidth are essential techno-vocabulary employed in professional, educational, and even social settings.

But those professionals have only recently been employing the
term "OT" — operational technology — to describe the systems
that connect IT systems with the real, physical world around them —
bringing electricity to their businesses, natural gas to their stovetops,
and water to their faucets.

"OT" is generally used to cover industrial control systems, supervisory control and data acquisition (SCADA) systems, programmable
logic controllers (PLCs), industrial sensors/transmitters, actuators,
and their supporting network communications [1,2].

The term may have arisen from the fact that leaders at industrial
firms often refer to the branch of the organization concerned with
operating the aforementioned systems as "operations", or the "operations side of the house". Thus, "OT" provides a convenient way to
emphasize a set of key differences between technologies.

1.1.   *Documentary evidence of an IT–OT cybersecurity gap*

To explore aspects of the term "operational technology" in academic
and professional literature, the authors performed a desktop analysis
and structured literature review using the IEEE Explore database.

The review began with a search for the exact text string "operational technology". Then, each resulting publication was manually
reviewed to determine (1) whether its usage referred to industrial
control systems components and communications (as per the description above); (2) whether the paper mentions differences between
IT and OT, and (3) whether the term's context was primarily
cybersecurity.

First, the results of the analysis found that the term "operational
technology" is coming into more common usage: Just seven papers
from 2014 used the term, while 24 used it in 2019. The vast majority of the papers from 2014 onward (82 of 88) used the term to
refer to industrial control systems and networks. Notably, the term
is used in *IEEE Std 1934–2018: IEEE Standard for Adoption of Open*

Table 1. Occurrences of "operational technology" in IEEE Explore database.

Year Published	Includes Term "Operational Technology"	Use Matches Definition	Mentions Differences between IT and OT	The Primary Focus Is Cybersecurity
1984–2013	11	0	0	0
2014	7	5	0	5
2015	7	5	1	1
2016	12	12	2	10
2017	20	19	6	7
2018	23	23	7	14
2019	25	24	13	20
Total	104	88	29	57

Fog Reference Architecture for Fog Computing, giving it some official status.

Second, the results showed that a smaller yet significant portion of the papers whose use matches the description above (29 of 88) mention differences between IT and OT.

Finally, the results showed that the term is commonly used within the context of cybersecurity. Fifty-seven of the 88 papers that use the term consistent with the proposed description use it within the context of cybersecurity (Table 1).

1.2. *Empirical evidence of an IT–OT cybersecurity gap*

In 2016, a leading U.S. industrial control systems integration firm invited author McBride to address a group of operations personnel from the firm's key clients. McBride discussed how the threat environment for industrial environments had evolved from the early 2000s, emphasizing how prevailing operational technologies were inherently vulnerable to cyberattacks due to inadequate consideration of abuse cases when the technologies were designed.

On the second day of the conference, the CEO of the integrator firm which had invited author McBride, recapped day 1, including the cybersecurity presentation and discussion. A refinery operator, who likely possessed the most life experience of anyone in the room, raised his hand and then explained in an annoyed tone of voice, "I appreciated everything about yesterday except the part about cybersecurity. I've been operating my refinery for 30 years. Never once has cybersecurity been an issue. I've been using the Modbus protocol for much of that time. It works exactly as intended. To me, cybersecurity is a self-fulfilling prophecy. The last thing I need is some cybersecurity guy from IT showing up to tell me how to do things. They will shut down my plant".

Other personal experiences and discussions the authors have had with cybersecurity consultants who work regularly in industrial environments confirm a common unfamiliarity, suspicion, and even distrust between those operating industrial facilities and IT groups.

1.3. *Description of the IT–OT gap*

Careful reflection on the results of the structured literature review together with the authors' experiences led the authors to attempt to characterize various aspects of the IT–OT gap, resulting in Table 2 "Aspects of the IT–OT Gap". Naturally, edge cases may not fit

Table 2.   Aspects of the IT–OT gap.

Aspect	IT	OT
Being controlled	Data	Physics
Measurement	Bits & bytes	Temperature, pressure, level, flow
Lifecycle	System lifecycle	Plant lifecycle
Consequences	Competitive disadvantage	Product damage
	Embarrassment	Loss of life
	Financial loss	Environmental release
Desired system	Confidentiality	Safety
characteristics	Integrity	Reliability
	Availability	Functionality
Educational	Computer Science	On the job
background of	Information Systems	Career & Technical Education
professionals	Cybersecurity	Electrical Engineering
Reporting chain	IT Manager	Shift Supervisor
	CISO	Plant Manager
	CIO	COO
Accounting	Cost center	Profit center

precisely within the aspects proposed, but the authors assert that he differences are significant and justify an intentional effort to overcome them.

1.4. *Problem statement*

Of particular interest is the aspect of the IT–OT gap entitled "Educational backgrounds of professionals". It seems logical to conclude that formalized training plays a role in creating the gap and therefore provides a significant opportunity to bridge the gap.

2. Research Approach

In thinking about how to effectively bridge the IT–OT cybersecurity gap, the authors contemplated the prevailing cybersecurity education paradigm within the United States. Since the early 1990s, schools have attested to the quality of their offering by seeking designation as a Center of Academic Excellence (CAE) in Cybersecurity from the U.S. National Security Agency [3]. Since 2014, those seeking such designation have had to show alignment between their curricula and official CAE "Knowledge Units" (KUs) [5].

Recognizing the importance of this paradigm, the authors determined to (1) review the existing CAE KU entitled "Industrial Control Systems and SCADA"; (2) if necessary, develop a list of categories for a replacement KU; and (3) create content for the proposed list of categories. The results of this research are presented in the sections below.

3. Analysis of Existing National Science Foundation Centers of Academic Excellence (NSA CAE) Industrial Control Systems Knowledge Unit

This section analyzes the 2020 Industrial Control Systems KU, found on page 64 of [4] — looking primarily at the KU's Intent, Outcomes, and Topics.

3.1. *Intent*

The intent statement is as follows:

> *The intent of an Industrial Control Systems Knowledge Unit is to provide students with an understanding of the basics of industrial control systems, where they are likely to be found, and the vulnerabilities they are likely to have.*

Analysis. The statement of intent seems to target a student whose primary role will not deal with industrial control systems — it provides basics and focuses on the "likely". One would expect that the outcomes which follow the statement of intent would align with the three named areas (basics, where they are found, and vulnerabilities) — but a careful review shows they do not.

3.2. *Outcomes*

To complete this KU, students should be able to do the following:

1. *Describe the use and application of PLCs in automation.*
2. *Describe the components and applications of industrial control systems.*
3. *Explain various control schemes and their differences.*
4. *Demonstrate the ability to understand, evaluate, and implement security functionality across an industrial network.*
5. *Understand and compare the basics of the most used protocols.*

Analysis. Outcomes 1–3 and 5 seem reasonable for a student who only needs peripheral awareness of industrial control systems — they lack specificity and do not address the differences associated with securing OT vs IT environments. Particularly concerning, given the statement of intent, is the absence of an outcome dealing with industries and processes which employ industrial control systems.

Outcome 4 merits individualized review. It is among the most complex and demanding of all objectives contained within the 2020 KUs: it requires a demonstration of understanding, evaluation, and implementation of security across a contextual space to which many universities have limited access; it seems to surpass the scope of the statement of intent and appears inconsistent with the nature of the other objectives within the same KU.

3.3. *Topics*

To complete this KU, all topics must be completed:

1. *SCADA Firewalls*
2. *Hardware Components*
3. *Programmable Logic Controllers (PLCs)*
4. *Protocols (MODBUS, PROFINET, DNP3, OPC, ICCP, SERIAL)*
5. *Networking (RS232/485, ZIGBEE, 900 MHz, Bluetooth, X.25)*
6. *Types of ICSs (e.g., power distribution systems, manufacturing)*
7. *Models of ICS Systems (time-driven vs event-driven)*
8. *Common Vulnerabilities in Critical Infrastructure Systems*
9. *Ladder Logic*

Analysis. These nine topics offer little intuitive categorization or prioritization versus other topics or terminology not on the list. For example, are SCADA firewalls more useful than non-SCADA firewalls for industrial networks? To what does the phrase "hardware components" refer? Why does the protocol list not include Ethernet/IP or HART? Does "Critical Infrastructure Systems" not merit a separate entry (outside the context of common vulnerabilities)? Is ladder logic a higher priority than function block logic?

In addition to a more intuitive structure, it would appear reasonable to expect case studies and specific ICS-related security guidance among the topics.

3.4. *Conclusion*

For the reasons described above, this analysis concludes that the 2020 NSA CAE KU lacks compelling organization and prioritization to effectively bridge the IT–OT cybersecurity gap.

4. Industrial Cybersecurity Knowledge Categories

In seeking to identify the knowledge categories that differentiate industrial cybersecurity from traditional cybersecurity, it would be necessary to involve a group of qualified participants in an appropriate research method.

The Idaho National Laboratory has over 200 individuals who focus exclusively on industrial cybersecurity, and provides a variety of services, including industrial cybersecurity training, for its government customers. When senior organization management heard about the need to develop improved guidance, they agreed to send 14 experienced professionals to participate.

4.1. *Nominal group technique*

The authors determined to use the nominal group technique to obtain and record expert input. The nominal group technique relies on anonymously written interaction among a medium-sized group of participants. This reduces the negative effects of personality, creates an accurate record, and diminishes coding bias. It allows the moderator and the participants to interact with one another for a robust treatment of the key question. Because written communication can occur among various parties simultaneously, it encourages broad participation within a relatively short timeframe. Van Den Ven and Delbecq report that the technique effectively elicits diverse perspectives [6–8].

The technique relies significantly on the moderator to carry the group through a decision-making process that achieves the objective. Before the session, the moderator reviews the objective of the session and prepares a rough script of the questions the group intends to address. The moderator uses a set of techniques, such as brainstorming, nominations, rankings, and voting to guide the process. As such, the moderator must have a strong understanding of the decision-making process, including approaches and options to give participants, know the software well, display general familiarity with the subject matter, and suspend his or her own bias.

It is important to note that in the nominal group technique the participants are not considered subjects of study but collaborators, anonymously imparting what they know to address a specific issue.

Simplot decision support center: The Simplot Decision Support Center (SDSC) is an in-person electronic meeting room located on the fourth floor of Idaho State University's Business Administration building. The facility features a 20-seat amphitheater, specifically designed to implement the nominal group technique for decision-making. The facility was used to develop NIST and CNSS training standards. The specialized software used in the Center produces an anonymous log of the input and records the decisions made by the group.

Participant qualifications: The group's professional background included titles such as Power Plant Operator, HVAC Specialist, Field Electrician, Information Security Technology Officer, Computer Technology Analyst – SCADA, ICS, and Cybersecurity Consultant. The group's former employers included Northern California Power Agency, Raytheon, National Security Agency, Virginia Transformer, El Paso Electric, and Phillips 66. In total, the group reported 31 years of experience in industrial cybersecurity, 32 years in non-ICS information security, and 88 years in industrial operations.

4.2. *Results of the nominal group session*

In response to the question, "What knowledge does an ICS security professional need to know that is not covered in standard information security?", the group identified 86 terms and concepts, which it organized into five categories:

- Control Knowledge
- Equipment
- Communications
- Regulations
- Instrumentation & Control

Analysis: Peer debriefing pointed out that the use of the word "control" in two of the categories could confuse users. The authors determined to replace the category "Control Knowledge" with "Industrial Processes & Operations".

Peer debriefing also pointed out that while safety was appropriately now within the Industrial Processes & Operations category, it was such a distinguishing feature between IT and OT that it should be in a unique category.

Additional reflection on the knowledge categories identified by the nominal group technique led the authors to recognize that the categories lacked a way to bridge the content "not normally covered in traditional cybersecurity" with what normally would be included, and the authors proposed three additional categories:

1. Common vulnerabilities
2. Defensive technologies and approaches
3. Events and incidents

5.   Content Creation for Knowledge Unit

The next challenge was to determine the content for each of the identified categories. As a starting point, the authors simply selected content based on their experience with industrial control systems. The authors then validated the results via critical comparison (triangulation) with external sources. The resulting NSA CAE-style KU is presented in the following.

5.1.   *Results*

5.1.1.   *Industrial Control Systems Knowledge Unit*

Intent: The intent of the Industrial Control Systems (ICS) KU is to ensure a foundational understanding of ICS, including their role in operating critical infrastructure, their key differences from information systems, their common vulnerabilities, and approaches to advancing their resilience.

Outcomes: Upon successful completion of this KU, participants should be able to do the following:

1. Describe industrial control systems, including the names and functions of their common components.
2. Identify several industry sectors and processes supported by industrial control systems.
3. Explain how industrial control system environments differ from information system environments.
4. Describe common weaknesses in industrial control system environments.
5. Describe approaches to address common weaknesses while considering unique ICS characteristics and requirements.

Topics: The following topics must be covered. Items in parentheses are examples of content and are strongly suggested.

- *Industrial processes and operations*: Industry sectors, professional roles, and responsibilities in industrial environments, engineering diagrams, process types, industrial life cycles
- *Instrumentation and control*: Sensing elements, control devices, programmable control devices, control paradigms, programming methods, process variables, data acquisition, supervisory control, alarms, engineering laptops/workstations, configurators, data historians
- *Equipment under control*: Motors, generators, pumps, compressors, valves, relays, generators, transformers, breakers, variable frequency drives
- *Industrial communications*: Reference architectures, industrial communications protocols, transmitter signals, Fieldbus systems
- *Safety*: Electrical safety, personal protective equipment, safety/hazards assessment, safety instrumented systems, lock-out tag-out, safe work procedures, failure modes
- *Common weaknesses*: Network architectures, unauthenticated protocols, outdated hardware and software, lack of training and

awareness among ICS-related personnel, transient devices, third-party access, supply chain

- *Events and incidents*: DHS Aurora, Stuxnet, Ukraine 2015, Ukraine 2016, Triton, Tam Sauk Dam, DC Metro Red Line, San Bruno
- *Defensive technologies and approaches*: Firewalls, data diodes, process data correlation, ICS network monitoring, cyber-informed engineering, security process hazards assessment, cyber-physical fail-safes, awareness, and training for ICS-related personnel

5.2.   *Validation*

5.2.1.   *Triangulation with Automation Competency Model*

To validate the topic contents for the first five knowledge categories, the contents were compared with the Automation Competency Model developed by the United States Department of Labor (DoL) with support from the International Society of Automation (ISA) [8].

As shown in Table 3, of the 38 terms provided as new topics, 30 are also found in the DoL model, representing a 79% match. Table 2 displays the locations of matches, which provide a useful resource for instructors seeking to use the proposed KU. It is noted that six of the seven terms missing a match are in the "Equipment under control category", which one might expect to find in the field of mechanical engineering or electrical engineering rather than industrial automation.

These should still be included in the topics list because this equipment directly influences the physical consequences of a cyberattack and cannot be ignored. The remaining term not found in the DoL Automation Competency Model is "electrical safety". Here, it is reasonable to assert that any cybersecurity professional who opens up a control enclosure to capture network traffic or update controller firmware requires basic awareness of electrical safety.

Table 3. Comparison of proposed knowledge unit topic terms with the Automation Industry Competency Model.

Knowledge Category	Term	Location in Automation Industry Competency Model		
Industrial	Industry sectors	p. 4		
processes and	Roles & responsibilities	3.2.1.1	5.6.19.3	
operations	Organizational roles			
	Engineering diagrams	5.2.14	5.3.13	5.5.13
	Process types	4.2.7	5.1.6	
	Industrial life cycle	4.1	4.1.6	4.1.7
Instrumentation	Sensing elements	5.2		
and control	Control devices	5.2		
	Programmable control devices	5.3.12		
	Control paradigms	5.3		
	Programming methods	5.3.17		
	Process variables	5.2.2		
	Data acquisition	5.7		
	Supervisory control	5.3.12		
	Alarms	5.5.7		
	Engineering computers	4.3.11.6		
	Configurators/calibrators	4.1.7.1	4.2.8.1	4.3.9.2
	Data historians	5.7.6		
Equipment	Motors	5.2.13		
under control	Pumps			
	Compressors			
	Valves	5.2.4	5.2.5	
	Relays			
	Generators	5.2.13		
	Transformers			
	Breakers			
	Variable frequency drives			
Communications	Reference architectures	5.6.1	4.2.9.2	
	Communications protocols	5.4.7	5.4.8	5.6.12.1
	Transmitter signals	5.2.6		
	Fieldbuses	5.4.7		
Safety	Electrical safety			
	Personal protective equipment	3.9.2.3		
	Safety/hazards assessment	4.5.5	4.5.11.3	
	Safety instrumented systems	5.5		
	Lock-out tag-out	4.5.11.4		
	Safe work procedures	4.5.11		
	Failure modes	5.5.8.3		

5.2.2.   *Triangulation with industrial cybersecurity guidance*

The content from the three additional topic areas (Common Weaknesses, Events & Incidents, and Defensive Technologies & Approaches) also requires validation. For the Common Weaknesses topic, the topics do not differ greatly from the content in traditional cybersecurity educational materials. But they may have unique implications for industrial environments. These topics are found in the publications given in Tables 4 and 5.

For the Defensive Technologies & Approaches category, some of the terms are found in existing industrial cybersecurity guidance. The unique characteristics of industrial environments make them of special interest. Independent sensing and backhaul, consequence-driven

Table 4.   Terms from the Common Weaknesses category and external location.

Term	External Location
Indefensible network architectures	NIST SP 800-82 p. C-6 [9]
Unauthenticated protocols	NIST SP 800-82 p. C-9
Unpatched, outdated hard & software	NIST SP 800-82 p. C-7
Lack of training and awareness	NIST SP 800-82 p. C-4
Transient devices	NERC CIP-003-8 pp. 24, 51–54 [10]
Third-party access	NERC CIP-013-1 pp. 3–4, 11–13 [11]
Unverified supply chain	NERC CIP-013-1 pp. 3–4, 11–13

Table 5.   Terms from the Defensive Techniques and Approaches category and external location.

Term	External Location
Firewalls	NIST SP 800-82 p. E-1 [8]
Data diodes	NIST SP 800-82 p. E-1
ICS network monitoring	NIST SP 800-82 p. E-1
Awareness and training for ICS-related personnel	NIST SP 800-82 pp. 4-1, 6-13, G-20
Cyber-physical fail-safes	NIST SP 800-82 pp. 5-21, G-64
Process data correlation	Academic papers [12–14]
Security process hazards assessment	Book [15]
Cyber-informed Engineering	Book [16,17]

cyber-informed engineering (CCE), and cyber process hazards assessment (Cyber PHA) are newer approaches especially applicable to industrial environments.

For the Events & Incidents knowledge category, triangulation may also provide validation; however, as avoiding and mitigating industrial cybersecurity events is the entire purpose of educating and training industrial cybersecurity professionals, such may hardly be necessary. A better question may be whether the chosen events are the most relevant or instructive of all possible events. While the investigation did not explore this question directly, it is possible to "reverse triangulate" the chosen events with the validated concepts to at least demonstrate that the events are consistent with the knowledge categories and items. Such is the approach described in the following section.

5.2.3. *Triangulation with industrial cybersecurity events and incidents*

As shown in Table 6, if an aspect of the attack, in the opinion of the authors, clearly involves the topic, it is marked with an X.

Of the 50 terms in the list, eight did not match an ICS-specific attack. Of the eight, three dealt with work safety, and as such were not relevant to this particular analysis. This left a 95% (45/47) match.

The eight unmatched items were control paradigms, motors, generators, pumps, compressors, electrical safety, personal protective equipment, and safe work procedures. While the relevance of these items may not have yet been proven by actual cyberattacks, all but one are covered by the three safety events also included in the events and incidents category, as illustrated in Table 7.

This mapping only left three items, control paradigms, electrical safety, and personal protective equipment, without validation by external documentation.

Table 6. Correlation of industrial cybersecurity-specific knowledge with industrial cybersecurity events.

		Industrial Cybersecurity Event			
		Stuxnet	Black Energy 3	Crash Override	Triton
Knowledge Category	Item	Iran 2009	Ukraine 2015	Ukraine 2016	Saudi Arabia 2017
Industrial	Sectors	X	X	X	X
processes and	Professional roles	X	X	X	
operations	Organizational roles	X			
	Engineering diagrams	X			
	Process types	X	X	X	X
	Industrial lifecycle	X	X	X	X
Instrumentation and control	Sensing elements				X
	Control devices	X		X	X
	Programmable control devices	X		X	X
	Control paradigms				
	Programming	X			X
	Process variables	X			X
	Data acquisition	X			
	Supervisory control	X	X		
	Alarms	X			X
	Engineering laptops	X			
	Configurators	X			
	Data historians	X			
Equipment under control	Motors				
	Generators				
	Pumps				
	Compressors				
	Valves	X			X
	Relays			X	
	Transformers		X	X	
	Breakers		X	X	
	VFDs	X			

(*Continued*)

Table 6. (*Continued*)

Knowledge Category	Item	Industrial Cybersecurity Event			
		Stuxnet Iran 2009	Black Energy 3 Ukraine 2015	Crash Override Ukraine 2016	Triton Saudi Arabia 2017
Communications	Reference architectures		X		X
	Comms protocols	X		X	X
	Transmitter signals	X			
	Fieldbuses	X			
Safety	Electrical safety PPE				
	Safety assessment	X			
	Safety systems (SIS)	X			X
	Lock-out tag-out				X
	Safe work procedures				
	Failure modes	X			
Common Weaknesses	Indefensible network architectures		X	X	X
	Unauth protocols	X		X	X
	Outdated			X	
	Lack of training		X		X
	Transient devices	X			
	Third-party access	X			
	Supply chain	X			
Defensive Technologies & Approaches	Firewalls				X
	Data diodes				X
	ICS network monitor	X		X	X
	Training		X	X	X
	CPS fail-safes	X			
	Process data	X			
	Security PHA	X			
	CCE	X			

Table 7.   Mapping of remaining items to other events.

Knowledge Item	Event			
	DHS Aurora	San Bruno	Tam Sauk	DC Metro
Control paradigms				
Motors				X
Generators	X			
Pumps			X	
Compressors		X		
Electrical safety				
Personal protective equipment				
Safe work procedures		X		

6.   Conclusion and Implications

In summary, this paper discussed the emergence of the term "operational technology" or "OT" to differentiate industrial control systems and related communications from "information technology" or "IT". The paper discussed various aspects of an IT–OT cybersecurity gap, focusing on the importance of formalized education to bridge the gap. The paper found prevailing ICS security curricular guidance ineffective at bridging the gap and proposed an updated KU, which used several techniques to validate.

It is foreseeable that this KU will be used to design or validate the content of a single course, or several modules within a course, taken by cybersecurity students. It is a solid starting point, yet insufficient to guide the creation of an entire industrial cybersecurity program.

Outcomes 3–5 (IT/OT differences, common weaknesses, unique defensive approaches) and Topics 6–9 (regulation, common weaknesses, events and incidents, defensive approaches) of the updated KU would be helpful (though certainly not sufficient) in developing industrial cybersecurity awareness, training, and education for individuals who already have an OT-related background.

References

[1] Gartner. Entry for "Operational Technology" in "Information Technology Glossary", n.d. https://www.gartner.com/en/information-technology/gloss ary/operational-technology-ot, last accessed 2022/12/03.

[2] National Institute of Standards and Technology. Entry for "operational technology" in Computer Security Resource Center Glossary, n.d. https://csrc. nist.gov/glossary/term/operational_technology, last accessed 2022/12/03.

[3] CAE in Cybersecurity Community. The National Centers of Academic Excellence in Cybersecurity (NCAE-C) Program's History, n.d. https://www.cae community.org/about-us/what-cae-cybersecurity, last accessed 2022/12/03.

[4] National Security Agency. 2020 Knowledge Units, n.d. https://www.iad. gov/NIETP/documents/Requirements/CAE-CD_2020_Knowledge_Units. pdf, last accessed 2022/12/03.

[5] Conklin, W.A., Cline, R.E., and Roosa, T. Re-engineering cybersecurity education in the US: An analysis of the critical factors. In *2014 47th Hawaii International Conference on System Sciences*, pp. 2006–2014, 2014. doi: 10.1109/HICSS.2014.254.

[6] Delbecq, A.L., Van de Ven, A.H., and Gustafson, D.H. Group techniques for program planning: A guide to nominal group and Delphi processes, Scott, Foresman, 1975. https://eduq.info/xmlui/handle/11515/11368.

[7] Van de Ven, A. and Delbecq, A. The effectiveness of Nominal, Delphi, and Interacting Group Decision-Making Processes. *Academy of Management Journal*, 17(4):605–621, 1974. https://doi-org.libpublic3.library.isu.edu/10. 2307/255641, last accessed 2020/06/11.

[8] U.S. Department of Labor, Automation Industry Competency Model V. 4., 2009 (updated 2018). https://www.careeronestop.org/CompetencyModel/ competency-models/pyramid-download.aspx?industry=automation,   last accessed 2020/06/05.

[9] Stouffer, K., Lightman, S., Pillitteri, V., and Abrams, M. Guide to Industrial Control Systems Security, 2015. https://doi.org/10.6028/NIST.SP.800-82r2, last accessed 2021/01/23.

[10] NERC. CIP-003-8 — Cyber Security — Security Management Controls, n.d. https://www.nerc.com/pa/Stand/Reliability%20Standards/CIP-003-8. pdf, last accessed 2021/01/22.

[11] NERC. CIP-013-1 – Cyber Security - Supply Chain Risk Management, n.d. https://www.nerc.com/pa/Stand/Reliability%20Standards/CIP-013-1. pdf, last accessed 2021/01/22.

[12] Krotofil, M., Larsen, J., and Gollmann, D. The process matters: Ensuring data veracity in cyber-physical systems, 2015. https://dl.acm.org/doi/10. 1145/2714576.2714599, last accessed 2021/01/22.

[13] Hadžiosmanović, C., Sommer, R., Zambon, E., and Hartel, P. Through the eye of the PLC: Semantic security monitoring for industrial processes, 2014. https://doi.org/10.1145/2664243.2664277, last accessed 2021/01/23.

[14] Ahmed, C., Zhou, J., and Mathur, A. Noise matters: Using sensor and process noise fingerprint to detect stealthy cyber attacks and authenticate sensors in CPS, 2018. https://doi.org/10.1145/3274694.3274748, last accessed 2021/01/23.

[15] Marszal, E. and McGlone, J. *Security PHA Review for Consequence-Based Cybersecurity*. International Society of Automation, 2019.

[16] Freeman, S., St Michel, C., Smith, R., and Assante, M. Consequence-driven cyber-informed engineering (CCE). United States, N. p., 2016. doi: 10.2172/1341416, last accessed 2021/01/23.

[17] Bochman, A. and Feeman, S. *Countering Cyber Sabotage: Introducing Consequence-Driven, Cyber-Informed Engineering (CCE)*. CRC Press, 2021. https://books.google.co.in/books?hl=en&lr=&id=2RkOEAAAQBAJ &oi=fnd&pg=PP1&dq=Countering+Cyber+Sabotage:+Introducing+14+ Consequence-Driven,&ots=4QdWXtzdiK&sig=OzKlYuRGDjwbJEMwFX 73ztQ7Jtk&redir_esc=y#v=onepage&q=Countering%20Cyber%20Sabotag e%3A%20Introducing%2014%20Consequence-Driven%2C&f=false.

Index

www.ingramcontent.com/pod-product-compliance
Lightning Source LLC
Chambersburg PA
CBHW050637190326
41458CB00008B/2316